To Gareth
from Dad

Happy Hunting!

Christmas 1989

CALIFORNIA MARINE LIFE

by

Marty Snyderman

DIVERS SU
50205
OF   ARBARA
19.9

First Printing, January 1988

ISBN 0-932248-07-1
Library of Congress Catalog Card Number 87-061515

Cover Design by James Graca
Illustrations by Anne Graca

MARCOR PUBLISHING, P.O. BOX 1072, Port Hueneme, CA 93041
Copyright © 1987 by Marty Snyderman

All Rights Reserved

Printed in Hong Kong

PHOTO CREDITS: All photographs taken by the author, with the exception of the following photographs credited by color plates as they appear in the book: Mike Farley (PLATE A-1/#1, PLATE D-4/#9 and #10); Fred Fischer (PLATE B-4/#15, PLATE C-2/#4, and PLATE C-8/#33); Steve Leatherwood (PLATE D-1/#2); Ron McPeak (PLATE A-2/#4); Tim Means (PLATE D-6/#13, PLATE D-7/#16 and #18, PLATE D-8/#20); Jeff Newman (PLATE D-7/#17); Steve Rosenberg (PLATE D-3/#8).

Table of Contents

Foreword

Have you ever returned from a trip to the seashore, a snorkel, or Scuba dive, a fishing trip or any other seagoing experience filled with questions about the animals and plants you have seen? I certainly have. Nearly 20 years ago, fresh from my basic scuba certification, I slipped over the stern of the *Scorpio II* into a kelp bed off Yellow Bluffs, Santa Cruz Island, for my first real ocean dive. For the next hour, I wound my way, bedazzled, through slivers of sunlight dancing in a forest of kelp, along steep rock faces studded with a myriad of shapes, colors, and textures, through a fish-filled cavern awash with surge, and across a brilliant flat of sun-painted sand. All around me in that gallery there were living things, many looking like nothing I had ever seen, or even imagined, before.

When I got back on shore, I searched through books to try and identify some of the life forms I had seen and to learn something about them. It is a habit I continue, many years and many dives later. Those searches for accurate and interesting information on marine life often have been time consuming and frustrating. There are scholarly texts, of course. But often they are so complex that only a serious scientitst can use them. And there are superficial, overly simplified booklets which usually show only a few of the most common, visible, or bizarre examples of life at sea. At worst these are inaccurate. At best they fall hopelessly short of satisfying a healthy curiosity. What I have wished for over the years is an authoritative text about marine life which is organized in such a way that it is easy to use. Happily, such a book now exists.

California Marine Life, by Marty Snyderman, presents the ocean and many of its inhabitants through the eyes of the beach walker, tidepool enthusiast, snorkeler, diver, fisherman, and marine naturalist. It groups the diverse life forms into categories that make sense to a casual observer and not just to a serious scientist, presenting them by the habitats in which they are most likely to be seen — mud flats and marshes, tidepools, sandy beaches, kelp forests, submerged reefs, sandy plains, and the unfathomable open sea. It features all the key players in each habitat and shares some important and interesting facts about each. Because of the logical way in which the guide is arranged, it is helpful both immediately after the experience and much later when the most vivid memories remaining are of the general surroundings (habitat) and a few of the most striking animals or plants encountered. Reading the treatment of that habitat will refresh your memory about other things you saw and will enlighten you about how creatures of the sea compete and cooperate in their complex world.

Marty Snyderman is a well-known author, photographer, cinematographer and naturalist. In *California Marine Life* he has written an entertaining and informative book. As an added bonus, he has lavishly illustrated it with extraordinary color photographs, mostly his own. The guide's organization makes it useful as a quick reference; its breadth of information makes it useful as an in depth source, not just for those exploring the ocean off California, but for their counterparts in many other temperate regions as well.

I think *California Marine Life* belongs on the book shelves and in the field bags of all of us who want to identify the animals and plants of the sea and marvel at how they go about their fascinating lives beneath the waves.

Stephen Leatherwood
San Diego, California

April 1987

Steve Leatherwood is the Senior Research Biologist at Sea World/Hubbs Research Institute in San Diego, a position he has held since 1978. An active researcher, Scuba diver, photographer and well-published author, the majority of Steve's work has focused upon the natural history of marine mammals. Steve is very well known and respected by his colleagues in the scientific community for the quality of his work, and he is also well known in the naturalist community for his ability to relate scientific information to those people who are primarily interested in practical applications in real world settings.

Acknowledgements

I would like to acknowledge and say "thank you" to several friends who have helped me throughout my diving career and especially with this book. That list includes Howard Hall, whose knowledge about the ocean and whose enthusiasm for diving and photography has been extremely helpful; Bob Johnson, who generously proofread this text; Gordie Heck, who as my boss enabled me to buy my first underwater camera system on credit, and a handful of diving buddies — Bob Cranston, Chip Matheson, Tom Allen, Mike Farley, Fred Fischer, Dan Walsh, Michele Hall, and Steve Earley — who I have had a lot of fun with.

I would like to acknowledge Bill Johnson, the owner/operator of the charter boats the *Sand Dollar* and the *Bottom Scratcher*, and Bob Cranston, owner/operator of the *Betsy M* for taking me out diving as many times as they have.

And I would especially like to thank my parents who encouraged me when I thought the light at the end of the tunnel was an oncoming train. I won't forget the lessons learned.

About the Author

Marty Snyderman is an assignment photographer, cinematographer, and author, specializing in the marine environment. While on assignment, Marty has enjoyed unique opportunities to film many sharks, whales, dolphins, manta rays, sea snakes, and many other large and unusual species in dive sites ranging from the Great Barrier Reef to the Bahamas. Having dived in California for more than a decade, Marty has an intimate knowledge about the marine wildlife found in state waters.

Marty's still photography and/or writing has been utilized by many noteworthy organizations and publications expressing interest in the marine environment including the National Geographic Society, the National Wildlife Federation, Sea World, Inc., the Monterey Bay Aquarium, Oceans magazine, Skin Diver magazine, Underwater USA, Nikon Inc., Diver Magazine, Ocean Realm, Newsweek, Scubapro Diving & Snorkeling, Aqua, California Diver, the San Diego Union, and many more. His cinematography has been seen on the National Geographic Society's television special entitled "Sharks," numerous episodes of Mutual of Omaha's "Wild Kingdom," NOVA's award-winning whale special, several PBS wildlife specials, Home Box Office, Pennzoil's adventure film series, "That's Incredible" "Ripley's Believe It or Not," and many more shows.

Marty has taught underwater photography seminars for Nikon Inc. and as a staff member for both PADI International College and the NASDS Diving Instructors College. Actively involved in the sport diving community in San Diego, where he resides, Marty works as a consultant in marine science with the San Diego Unified School District, and he continues to teach seminars in underwater photography through several dive stores.

Introduction

The first time I ever saw a California gray whale underwater I was diving in 60 feet of water off the coast of San Diego in the La Jolla submarine canyon. This canyon lies near the southernmost extension of a series of deep marine trenches that parallel the coast of Southern California. Earlier in my dive, I had been preoccupied by my efforts to film a tiny, yet magnificently colored purple and orange nudibranch that is commonly called the Spanish shawl and that is referred to by members of the scientific community as *Flabellinopsis iodinea*. Nudibranchs are often described as sea slugs or shell-less snails in textbooks, but their beautiful coloration makes them virtually indescribable without an accompanying photograph.

I don't know why I looked up at the precise moment that I did. I'd like to say I have some 6th sense about me when I am in the water, but more likely I just looked up because I was out of film. Whatever the reason, I am sure glad I did. The whale was a beautiful sight, creating an image in my mind's eye that I will treasure throughout my lifetime.

At first I was somewhat overwhelmed by the whale's size. Although gray whales are relatively small whales, they have body parts—the head, pectoral flippers, and tail—that are much larger than a fully grown human being. The great creature was obviously powerful, yet graceful, aware of my presence I think, and an absolute joy to observe. I remained motionless and just stared. I wasn't frightened and certainly never felt threatened, but I was in awe of the sight before me. The whale gave me the once over with its tennis ball sized eyes and then disappeared within a couple of seconds.

It was one of the few times in my diving career I was glad I couldn't take a photograph. My camera would have been a burden, robbing me of the pure pleasure of watching, of just being there next to a gray whale. I know how few people ever swim eyeball to eyeball with a whale in the wilderness, and I will remain forever grateful for that opportunity.

When I reflect upon my California diving experiences I often think back upon that moment. Somehow, symbolically that dive capsulizes my experiences in California better than anything else. For in California we do indeed have the opportunity to enjoy an incredibly diverse range of marine wildlife that varies from inch long, rainbow colored nudibranchs to 40 foot long, 40 ton California gray whales.

Being able to enjoy this incredible variety of marine plants and animals in the majestic settings of the Pacific provides a wonderful form of relaxation and recreation for many Californians. For there is a striking contrast in California between the overly populated metropolitan areas whose motto appears to be "life in the fast lane" and the almost untarnished frontier of the Pacific Ocean. In coastal waters, often less than 100 yards offshore and less than a few miles from the heart of Los Angeles, San Diego, Santa Barbara, and San Francisco, whale watchers, birders, boaters, fishermen, snorkelers, and divers can escape into a marine wilderness that is rich in sea life and adventure. In a state that has an international reputation for being one of the most technologically advanced areas in the world, complete with overly demanding, stress filled "on the go" lifestyles, spending a day on the Pacific provides a perfect escape from the harsh realities of urban pressures. Those of us who are fortunate enough to live in or visit California know the state is truly blessed with the opportunity to enjoy a marine wilderness that is as magnificent and diverse as any wilderness area on earth.

About the Book

Many people first venture into the marine environment while participating in recreational activities . . . snorkeling, Scubadiving, boating, swimming, fishing, beachcombing, and sight seeing to name a few. Others are drawn to the sea just to relax and get away from the hassles of every day life. Whatever the reason, for most people the common ingredients are the pursuit of fun, knowledge, and a desire to enjoy the wonders of nature. With that thought in mind, I have written this book as both a personal guide and reference source for recreational enthusiasts.

Although I have included some basic scientific terminology, this book is not intended strictly as a scientific text. Instead, the text, photographs, and illustrations are designed to help you acquire a practical knowledge of the natural history of the plants and animals found in California's marine environment. After all, many of us are primarily interested in learning when and where we are most likely to be able to observe various plants and animals; who is closely related to whom, and why; and who eats what, and who likes to eat whom. In addition, we enjoy understanding some of the more astonishing adaptations used by various species to ensure their survival.

This type of knowledge helps all of us better understand why in California we are more likely, for example, to see a gray whale in the winter, a lobster in a crevice in a rocky reef, a halibut in the sand, a curlew on a sandy beach, and not very likely to see a blue shark in a kelp forest community close to shore. The bottom line is . . . the more we know about the environment we are in, the richer the setting seems to be.

How to Use This Book

This book is designed to be used in two ways. First, it can be used as a reference source which provides an in-depth discussion of the natural history of the plants and animals found within California's marine kingdom. If you desire to utilize it in this manner, the best thing to do is to read the book from cover to cover. In doing so, you will receive the maximum benefit. Second, this book is intended to assist you in a "real world" setting in the wilderness, and can be used quite easily to help you become familiar with the wildlife that you observe. As an example, if you see a particular plant or animal, and you want to learn some basic information about its natural history, that information is readily at hand.

If you know the organism's common name and want to learn its scientific name, refer to the "Cross Reference Index from Common Name to Scientific Name" which begins on page 241. The index is arranged in alphabetical order.

The plants and animals discussed in the text are grouped according to the habitat (**the beach, the kelp forest, the rocky reef, the sand, and the open sea**) in which they are most likely to be observed by recreational enthusiasts. So if you do not know either the common name or the scientific description, your best bet is to scan the identification photographs that appear in the color sections throughout the book and which are grouped according to each habitat. Caption information for each photograph will provide the common name of the species that is pictured. The common name can be cross referenced with the scientific name by using the proper index.

As you are probably aware, some animals are seen in more than one habitat over the course of their lives. The corresponding illustrations or identification photographs for those creatures are placed within the habitat section in which the animal is most often observed. So if you do not find a plant or animal pictured in the habitat where you first encountered it, flipping through the photographs and illustrations within the other habitat chapters might prove beneficial.

In the case of both **marine mammals** and **marine birds**, you will find corresponding identification photographs in the color sections entitled "California Marine Mammals," and "California Marine Birds."

If you find pictures of animals that strongly resemble the species you are researching, but you are not positive about the exact identification, keep the following facts in mind: (1) many animals experience different color phases during different periods of their lives, and (2) the appearance of males and females of the same species often varies dramatically, as do the markings of juveniles and adults. And not every species that has ever been reported to inhabit California's marine kingdom is pictured. Instead, I have featured the plants and animals most often encountered by recreational enthusiasts. This feature helps make the book an easy to use field guide and quick reference source.

So in some cases, you might see animals that bear strong resemblance to those pictured in the various identification sections, but you believe the animal you saw and the one in the picture to be slightly different. If that is the case, you can probably draw some general conclusions about the natural history of the animal you are researching by reading about similar animals. In doing so, be sure to refer to the phylum overviews in the chapter entitled "Classifying Plants and Animals: (A Quick Review)." The brief phylum overviews which begin on page 18 provide excellent general information about the common traits shared by all species that are grouped within a given phylum.

Delphinus delphi

Pisaster giganteus

Larus californicus

Orcinus orca

Charcharodon carcharius

Hypsypops rubicundus

Macrocystis pyrifera

Classifying Plants
and Animals

Thalasseus elegans

Flabellinopsis iodinea

Xiphias gladius

California gull

Killer whale

Knobby star

Giant kelp

Great white shark

Garibaldi

Swordfish

Harbor seal

Common dolphin

Classifying Plants and Animals

Many of us remember the basics from our high school biology classes, being aware that plants and animals are given scientific names called their genus and species. Those names were the long Latin words that we had to memorize even though the only living thing that knew those names, or that seemed to care, was the teacher. At the time, memorizing those names seemed more like another meaningless exercise in discipline, rather than something which would provide a useful foundation for communicating about plants and animals.

Obviously, for the exactness required within the scientific community, the use of scientific names is critical in order to prevent even the slightest misunderstandings. The scientific community must be able to distinguish each species from every other species on our planet. But it is also important to realize that even for many people who are not members of the scientific community, the use of scientific nomenclature often proves valuable. This is because the use of common names alone can often lead to confusion about the particular species in question.

Not only is there a distinct lack of standardization of common names, but many species have several common names. For example, a commonly seen starfish in California waters is scientifically named *Pisaster giganteus,* but in lay terms this creature is named the knobby starfish, the giant-spined star, the giant star, and the giant sea star. To further compound the problem of using common names, the same common names are used in different areas to describe totally different species. As an example, the name "red crab" is often used to describe two entirely different animals, one of which is commonly found in reef communities in Northern California and is highly valued as a commercial food source, while the other is much smaller, of no commercial value, and only occasionally seen near shore. The smaller species is also referred to as a pelagic red crab, a tuna crab, and a squat lobster.

Thus, it is easy to understand the benefits gained by using scientific terminology. But the trouble is, if you are like many people, you have long since forgotten much of the information you once had to memorize. If you fit into this group, but you really would like to be able to recall at least some of the basics, the information in the remainder of this chapter will prove especially helpful. It will enable you to identify specimens and gain a rudimentary knowledge about most of the organisms you are likely to encounter in California's marine wilderness. In many cases positive identification by species will be possible. This is especially true of those animals that are most often encountered by people who spend time in the marine environment.

Scientific Classification System: A Quick Review

Nearly 1.5 million currently living species of plants and animals have been described by the scientific community. In order to study these life forms, scientists have grouped them into various categories according to commonly shared traits. This is called taxonomic classification. The first major division is that of kingdoms, the kingdom of plants and the kingdom of animals.

Many species of plants play an important role in the ecology of California waters, but most people exhibit strong interest only in the larger plants that we call kelp. Smaller marine plants rarely capture their attention. Therefore, I have included a brief description about California's marine plants at the end of the section entitled "Life in the Kelp Forests," but have chosen to devote the vast majority of space to the animals found in California's marine kingdom.

The Animal Kingdom

Animals are classified, or grouped, according to the following heirarchy:

> Phylum
> Class
> Order
> Family
> Genus
> species.

The Significance of Bilateral Symmetry

The phenomenon of bilateral symmetry is of fundamental significance to those who adhere to the theory of evolution. Animals that display bilateral symmetry — as opposed to animals that display radial symmetry — possess a right half and left half that are essentially mirror images of one another. Bilaterally symmetrical animals like fish also have a distinct top and bottom, and distinct posterior and anterior ends.

Bilateral symmetry allows for the specialization of different parts of the body in relation to different functions. As a result, bilaterally symmetrical animals can move from place to place more efficiently than animals that are radially symmetrical. Animals that display radial symmetry, such as sea stars, are generally sedentary, moving only when they absolutely have to for the sake of survival. Animals that are bilaterally symmetrical are far better at locomotion, seeking food, finding mates, and in avoiding predators.

Rather early in the evolution of bilateral symmetry, specialized organs and tissues that are important in capturing food, avoiding predators, and in monitoring the environment were grouped toward the head. This trend is of fundamental importance in the evolution of animals. Structures that dealt with other bodily functions were located farther back in the body. Over the course of time the number and complexity of specialized organs in animals with bilateral symmetry continued to increase. The central nervous system of these animals became concentrated large longitudinal nerve cords. Nerve cells also became grouped toward the head, an evolutionary process which eventually led to the development of the head and brain found in specimens of the more advanced phyla.

A specific organism is named according to its genus and species. When correctly printed in a text, the genus and species are italicized, and the genus is always capitalized while the species is not.

Organisms are also classified in a broader sense by the just listed categories. The chart below provides the taxonomic, or scientific classification of the species we normally refer to as the California spiny lobster, the garibaldi, and the California gray whale.

Common name	California spiny lobster	garibaldi	California gray whale
Kingdom	Animalia	Animalia	Animalia
Phylum	Arthropoda	Chordata	Chordata
Class	Crustacea	Osteichthyes	Mammalia
Subclass:	Malacostraca		
Order	Decopoda	Perciformes	Mysticeti
Family	Palinuridae	Pomacentridae	Eschrichtiidae
Genus	*Panulirus*	*Hypsypops*	*Eschrichtius*
species	*interruptus*	*rubicundus*	*robustus*

The scientific task is to describe an animal by starting with the phylum, and then work down to a specific genus and species. The animal kingdom consists of approximately 35 phyla, so many species of animals are grouped in the same phylum. The phylum is subdivided into classes, which in turn are subdivided into orders, the orders into families, the families into genera, and finally the genera are divided into individual species. As you'll note in the chart above in the classification of California spiny lobster, some of the categories are further divided into subcategories such as subclass, but such divisions are beyond the scope of this text.

The specific categories have been created by scientists out of both necessity and convenience. They are arranged to reflect known or assumed evolutionary relationships among animals. If an animal has a characteristic that is not commonly shared with others in a broad group, it is important to realize the animal is not imperfect, the man made system is.

It is obvious in a scientific, or evolutionary sense, that garibaldi and gray whales are more closely related to each other than they are to spiny lobster. All three are in the Kingdom Animalia, but spiny lobster are in the Phylum Arthropoda, a phylum which describes many invertebrates, animals that lack a spinal cord. Garibaldi and gray whales are vertebrates, grouped together in the Phylum Chordata. It follows that one can logically assume that humpback whales are more closely related to gray whales and garibaldi than they are to spiny lobster.

Certain generalizations can be made about a given species by knowing its Genus, or its Family, or its Class etc., and you will often be satisfied just by knowing more or less where an animal fits into nature's grand scheme. That seems especially true with any number of rather small and seemingly inconspicuous invertebrates such as many of the worms, crabs, and snails.

Phylum by Phylum Overview

The following overviews of the scientific phyla (the correct plural of the word phylum) will begin with the simplest of microscopic life forms, protozoa, and work toward the most highly specialized and complex forms, the marine mammals. Organizing the phyla strictly by complexity is difficult, since the scientific community is still debating the proper order. The important point to remember is that plants and animals within the same phylum share common physiological characteristics.

Protozoa
plankton

This phylum includes a diverse variety of microscopic animals. The unifying common characteristic is that all are one celled animals. Most protozoans are benthic, attaching themselves to plants or rocks, but some forms are pelagic, meaning they live up in the water column. While protozoans play a vital role in most food chains, non-scientists rarely pay much attention to their presence because of their minute size, except in the case of observing the light shows exhibited by bioluminescent forms.

Many, but not all, forms of both phytoplankton (plant plankton) and zooplankton (animal plankton) are included in this phylum. Some organisms referred to as plankton even within the scientific community

Classifying Marine Animals

Scientific Phylum	Common Name
Protozoa	plankton
Porifera	sponges
Cnidaria	corals, hydroids, sea anemones, jellyfish, by-the-wind sailors, sea fans, sea pens, sea pansies
Ctenophora	comb jellies
Ectoprocta	bryozoans
Platyhelminthes	flatworms
Annelida	plume worms
Mollusca	abalone, nudibranchs, octopi, squid, scallops, sea hare, mussels, oysters, clams, isopods, periwinkles, chitons, snails, limpets
Arthropoda	barnacles, shrimp, crabs, lobster, amphipods, copepods
Echinodermata	sea stars (starfish), brittle stars, sea urchins, sand dollars, sea cucumbers
Chordata (vertebrates)	tunicates, salps, sharks, rays, skates, all bony fishes including eels, sea lions, seals, whales, sea otters, dolphins and birds

include the larval stages of various crabs, lobsters, scallops, and the juvenile stage of fishes, animals that are multicellular and obviously are not protozoans. In some discussions it is important to note whether the term plankton refers to single celled phytoplankton and zooplankton, or whether the grouping includes the earlier life stages of more complex animals.

In recent years some scientists have began to support the theory that all animals are multicellular, thus excluding all protozoans from the animal kingdom and placing them into a newly named kingdom called Protista, but this theory is still being debated.

Porifera
sponges

The phylum Porifera consists of approximately 10,000 species of marine sponges. Sponges are among the simplest and most primitive of multicellular animals. The cells lack specialization and are only loosely associated, thus lacking definite symmetry. Sponges do not possess specialized tissues and organs. Many sponges are colonial, some species are encrusting, and still others are erect and lobed. As adults all sponges are sessile, being attached to the bottom, but in their larval stage sponges are free-swimming.

Sponges have a unique system of feeding which involves the flow of nutrient rich water through a system of small pores found throughout the body. Also used for respiration, the pores are lined with flagella which force the water through the openings by rhythmic beating. The name of the phylum Porifera is derived from the numerous pores.

Cnidaria

(previously grouped in the phylum named Coelenterata, which has now been divided into two phyla, Cnidaria and Ctenophora)

corals, hydroids, sea anemones, jellyfish, by-the-wind sailors, sea fans, sea pens, sea pansies

The phylum Cnidaria (pronounced with a silent "C") describes a wide group of relatively simple animals including corals, hydroids, sea anemones, jellyfish, by-the-wind-sailors, sea fans, sea pens, and sea pansies. The group is characterized by their radial symmetry. All specimens possess a centrally located mouth and all utilize stinging cells called cnidocytes located in their tentacles to help capture prey.

Inside the cnidocytes there is a nematocyst, a harpoon-like structure which resembles a coiled, thread-like tube that is lined with a series of sharp barbs. Nematocysts are found only in animals in the phylum Cnidaria. Cnidarians use the spear to snare their prey, and then they gather in the harpooned victim with their tentacles. The nematocyst is stimulated to fire by both chemicals and touch. Using water pressure to discharge, the firing of a nematocyst is one of the fastest cellular processes in all of nature, occuring within a time span of approximately 3 milliseconds and at a rate of speed of 2 meters per second.

Due to the high speed attained by the filament of the nematocyst the barb is capable of penetrating through the hard shells of many crustaceans. Once penetration is achieved toxic chemicals are released to stun the prey and then additional tentacles are used to draw the prey into the mouth. Cnidocytes can discharge only once and then they are absorbed into the body and a new stinging cell is formed. Some adult forms such as jellyfish are free-swimming and are referred to as medusae, while the benthic forms are often called polyps. In the polyps the animal's mouth faces away from the substrate to which it is attached, while in the medusae the mouths generally face down and are surrounded by downward hanging tentacles.

In past years the members of the phylum Cnidaria and the phylum Ctenophora were grouped together in a phylum named Coelenterata, but in recent times the two groups have been separated. This is due to the fact that unlike members of the phylum Cnidaria, ctenophores lack stinging nematocysts, and the bodies of ctenophores are more complex than those of cnidarians.

Ctenophora
comb jellies

This phylum consists of about 90 species, all of which are marine animals. Most ctenophores are commonly called comb jellies. Boaters and even divers rarely pay much attention to any specimens because most are nearly transparent and less than a few centimeters in size. Ctenophores are capable of self propulsion, manuevering with the use of wavelike movements from the bands of cilia called ctenes. Their gelatinous bodies display radial symmetry. Fossil remains discovered in West Germany in 1983 indicate that the basic body plan of comb jellies has remained unchanged for close to 400 million years.

Ectoprocta
bryozoans

Although this phylum includes over 4,000 marine species, it is not well understood. The animals in this group are generally referred to as bryozoans. Many species are quite small and require the use of a microscope to differentiate between them. Most species are colonial and all attach to the sand, to rocks, to kelp, or to other animals.

The individuals making up a colony are called zooids. The zooids consist of an elongated body topped by a circle of tentacles that surround the mouth. Zooids lack specialized organs for respiration, excretion, and circulation. To compensate, the zooids form a hard, permanent outer shell which overlays a thicker layer of calcium carbonate and an interior cavity that contains a digestive tract. Cilia on the inner surface of the ring of tentacles assist in moving captured food to the mouth.

Colonies of zooids form either 1) flat encrusting sheets, such as those found on the blades of giant kelp plants, or 2) upright colonies which resemble lace-like flower petals, and that are sometimes found on rock, sand, or clay bottoms.

Platyhelminthes
flatworms

The most prominent members of this phylum are flatworms. They are to be distinguished from the plume worms which are in the phylum immediately following, called Annelida. Unlike their land dwelling cousins, many flatworms are quite striking from a visual perspective. Flatworms are just that, flat. Most are both bottom dwelling and parasitic. The nervous system of flatworms is very simple, although there are definite longitudinal nerve cords which comprise a simplistic central nervous system. Light sensitive cells in their "eyespots" enable flatworms to distinguish light from dark, and flatworms tend to avoid or move away from areas that are strongly lit. Locomotion is accomplished primarily by use of hair-like cilia which is most highly developed on the underside of the body. A flatworm has a definite front and rear end, right and left side, and top and bottom. The mouth is centrally located.

Flatworms are far more active than radially symmetrical animals such as cnidarians and ctenophores. This level of increased activity corresponds to the greater concentration of sensory nerves toward the head.

Annelida
plume worms

It is a confusing fact to many beginning biologists that plume worms are members of this phylum while flatworms are included in an entirely different phylum, Platyhelminthes. Plume worms, sometimes called Polychaete worms, are segmented, and each segment possesses a pair of paddle-like extensions called parapodia which assist in locomotion. This group includes featherduster worms which divers usually spot by the presence of their feather-like gill plumes.

Mollusca
abalone, nudibranchs, octopi, squid, scallops, sea hares, mussels, clams, oysters, isopods, periwinkles, chitons, snails, and limpets

Mollusks are such a large group that characterizing them proves difficult. Species of mollusks include abalone, nudibranchs, octopi, squid, scallops, sea hares, mussels, clams, oysters, isopods, periwinkles, chitons, snails, and limpets. In general, the common unifying features are that mollusks are unsegmented, and have well-developed sensory organs which are concentrated toward or in the head. Most, but certainly not all, have a hard external shell, and many have a large muscular organ called a foot which is used for locomotion, anchorage and securing food. A commonly shared characteristic of most mollusks is a radula, which is a rasping, tongue-like organ used to secure food. The structure of the radula is species specific, and is found in all mollusks except bivalves — mussels, scallops and oysters.

The phylum Mollusca is comprised of 7 classes, 4 of which have specimens that are quite commonly observed in California waters. The class Amphineura includes the chitons which are generally found in rocky intertidal areas. The class Gastropoda (Gastropods) includes snails, slugs, sea hares, limpets, abalone, and nudibranchs. The class Pelecypoda contains the bivalves, including mussels, clams, oysters, and scallops. Though free swimming in the larval stage, most are benthic as adults, and some are cemented to the substrate.

The most highly evolved class of mollusks is the group called Cephalopoda (head-foot animals), which includes squid and octopi. These animals are characterized by their sucker-lined tentacles, advanced sensory organs, and reduced or lack of external shell.

23

Arthropoda
barnacles, shrimp, crabs, lobsters, amphipods, copepods

Members of the phylum Arthropoda account for more than 75 percent of all living animals. Most are insects and are, therefore, not considered to be marine life forms. The most prominent common feature of arthropods is a protective exoskelton that is usually jointed. The hard skeleton provides protection and gives muscles a place to attach. Periodically the exoskeleton is shed, or molted, so that growth is possible. When molting, arthropods are especially vulnerable to predators, and the new exoskeleton is quickly formed.

The class Crustacea (crustaceans) is found in the phylum Arthropoda and includes all shrimp, lobster, crabs, as well as barnacles, copepods, and amphipids. In addition to other common characteristics, most members of this class have a pair of antennae.

Echinodermata
sea stars (starfish), brittle stars, sea urchins, sand dollars, and sea cucumbers

Echinoderms are an ancient group of marine animals that are very well represented in the fossil record. Echinoderms are found only in the marine environment. Today there are approximately 6000 species of echinoderms living in the oceans of planet Earth. The phylum of Echinodermata includes sea stars (starfish), brittle stars, sea urchins, sand dollars, and sea cucumbers, as well as the crinoids so often seen in tropical waters. In recent years, in some educational circles the term sea star is replacing the word starfish to emphasize that these creatures are not true fish. In this text, the two terms are used interchangeably.

Echinoderms are characterized by a calcerous skeleton, external spines or knobs, and five-sided, or pentamerous, symmetry. In larval stages many species display bilateral symmetry. As adults, a specialized system of tube feet is controlled by a water vascular system. The feet are not only locomotive organs, but also serve respiratory, excretory, and sensory functions. The presence of tube feet as part of a water vascular system is unique to this phylum. As adults, echinoderms have no head or brain.

There are 5 classes of echinoderms, 4 of which are represented in California. Members of the class Echinoidea — which describes all sea urchins, heart urchins, and sand dollars — are either sediment ingestors and herbivores. Sea stars, on the other hand, are generally carnivores and are grouped together in the class Asteroidea. Sea stars are among the more important predators in many marine biomes. The class does not include the brittle stars which are in the class Ophiuroidea. Some brittle stars prefer to sift through the substrate ingesting a variety of organic particles, while others capture microplankton suspended in the water column.

Sea cucumbers are sausage-shaped echinoderms in the class Holothuroidea. Sea cucumbers have a mouth at one end that is surrounded by modified tube feet, a muscular body wall with reduced skeletal units and fewer spines than many other echinoderms. They are generally described as sediment feeders.

The Vertebrates
tunicates, salps, sharks, rays, skates, all bony fishes including eels; sea lions, seals, whales, sea otters, dolphins, and birds

Worldwide the number of species of chordates, or vertebrates, numbers approximately 45,000. The phylum Chordata obviously includes all fish, amphibians, birds, reptiles, and mammals. Surprising to many, it also describes all benthic sea squirts, tunicates, and planktonic gelatinous salps. To be classified as a vertebrate, an animal needs to have at some stage of development: 1) a supportive, flexible notochord made of cartilage, 2) a single, hollow dorsal nerve cord, and 3) pharyngeal slits, or openings, which develop into gill slits in the case of fishes. Pharyngeal slits are present in the embryos of all vertebrates, but are lost in the latter stages of development of terrestrial vertebrates. Sea squirts, tunicates, and salps possess a notochord and nerve chord only in their larval stages.

In addition to the 3 just mentioned features, chordates have several other features which tend to distinguish them from other phyla. All are more or less segmented, and they have an internal skeleton against which muscles can work. Compared to other animals, the chordates demonstrate remarkable powers of locomotion.

The sea squirts, tunicates, and salps belong to the subphylum Urochordata. These animals lack the distinct skull found in other vertebrates. Sea squirts can be either solitary or colonial, but California divers most often encounter solitary forms. One solitary sea squirt commonly called the stalked tunicate (*Styela montereyensis*) looks like the pod of a terrestrial plant. Reaching a height of 8 to 10 inches, the animal is connected to the substrate by a long, thin stalk. Stalked tunicates feed via ciliary acton of an incurrent siphon, and rid themselves of waste through an excurrent siphon. Both siphons are located on the tip of the pod-like body.

Those animals that possess a vertebral column which extends through the main axis of the body for support are in the subphylum Vertebrata. This grouping includes all fishes, mammals, and birds. There are more species of fish in the world than all the rest of the vertebrates added together. Fishes are subdivided into 3 classes, two of which are rather significant. The class Chondricthyes is composed of sharks, skates, and rays, those fish that have cartilaginous skeletons. The class Oesteichthyes describes all bony fishes. Fishes with skeletons evolved later and are far more numerous.

All marine mammals belong to the class Mammalia. This class includes a variety of species of sea lions and seals which are collectively called pinnipeds, as well as sea otters, whales, and dolphins. In addition to being air breathers, mature females in all species of mammals have mammary glands that are used to nurse their young. Mammals are warm-blooded, have a four-chambered heart, advanced locomotion compared to reptiles and amphibians, and specialized teeth. Most mammals bear live young.

Other air breathing vertebrates are represented by the class Reptilia, the turtles, which are only rarely encountered in California, and the class Aves, the Birds. I have also chosen to present a special segment on marine birds, because they play such vital roles in the ecology of the sea. This section begins on page 211.

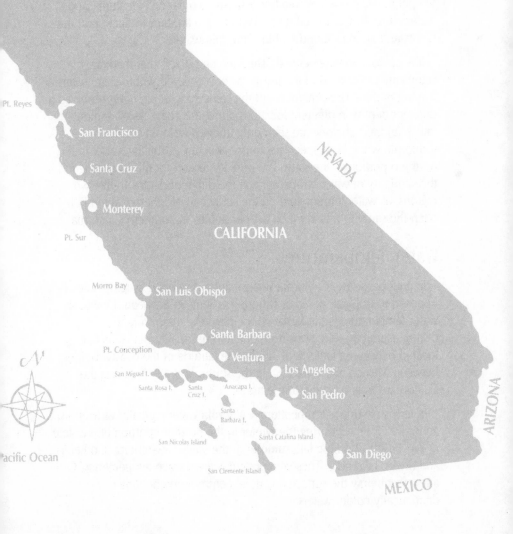

California:
The Big Picture

OREGON

ureka

Pt. Reyes

San Francisco

NEVADA

Santa Cruz

Monterey

CALIFORNIA

Pt. Sur

Morro Bay

San Luis Obispo

Santa Barbara

Pt. Conception

Ventura

San Miguel I.

Los Angeles

Santa Rosa I. Santa   Anacapa I.
              Cruz I.

San Pedro

ARIZONA

Santa
Barbara I.

San Nicolas Island

Santa Catalina Island

San Diego

Pacific Ocean

San Clemente Island

MEXICO

California: The Big Picture

In acquiring an understanding of the life forms that inhabit California's marine kingdom, it is of fundamental importance to consider water temperatures, prevailing weather patterns, geographical features, and water currents. Combined, these factors have a major impact upon the plants and animals found within California water.

It is obvious to experienced fishermen and divers that there are significant differences in the life forms encountered within state waters at varying depths. Thus, members of the same species are often observed at different depths in different locations along the coast. As examples, ratfish and red abalone are seen only in deep water in Southern California while both species commonly occur in shallow water in the northern portion of the state. The primary reason for this phenomenon is the similarity of water temperature at the different depths in the two regions, as water temperature on the surface in Northern California is often the same as it is at much deeper depths in Southern California.

Water Temperatures

In a global sense, California water can be described as temperate, as opposed to tropical, polar, or various gradients in between. Obviously water temperature is the factor which plays the major role in distinguishing these regions from one another. California waters are located between a latitude of 32° N, the latitude of the coastal border between Mexico and California in the south, and 42° N, along the California/Oregon border to the north.

In terms of surface temperature, California water typically varies from about 49° F (9° C) during the winter in the northern portion of the state to 70° F (21° C) during late summer in the south. Swimmers, snorkelers, and divers, take note. These temperatures are surface temperatures. Only a few feet below the surface it is quite common to encounter dramatically colder waters.

Generally speaking, Pt. Conception is a point of demarcation in terms of water temperatures within the state. Pt. Conception is not a geographical midpoint, being located well south of the center of California's Pacific coast. But due to the lay of the land along the coast and the prevailing currents and weather patterns to the north and south, conditions above and below this outcropping of land vary considerably. North of the point, water conditions are usually considerably rougher and temperatures are normally 4 to 10 degrees F cooler than to the south.

Mixed layers of warm and cold water throughout the water column are the rule, not the exception. The dividing line between bands of water of differing temperatures is called a thermocline. If not for surface winds, water temperature would uniformly drop with increased depth. Snorkelers and Scuba divers would encounter warmer water near the surface due to the warming effect of the sun's rays and colder water as they descended. However, winds along the sea's surface create water movement which in turn causes various layers of water of different temperatures to mix. To divers the most obvious sign of a thermocline is the radical change in temperature. A drop of 5 degrees F or more is quite common, and 10 degrees F is not unheard of. Thermoclines are more pronounced in Southern California.

A less immediately obvious effect, but one that is certainly as interesting, is that you'll often discover many species of marine life showing a distinct preference for specific ranges of water temperatures. While it is true that the same life form may occur in a wide range of water temperatures, in some settings it is common to see sharply delineated lines of fish or any number of invertebrates holding their position just above or below the thermocline. Noting those demarcations enables us to better understand the importance of water temperature upon the lives of sea creatures.

Prevailing Currents

California waters are significantly impacted by the California Current, a major surface current system that brings cold water from the Gulf of Alaska. The flow is usually described as slow, though the intensity will vary with the seasons. During late summer and fall the current tends to hug the coastline of Northern and Central California more closely than in winter and spring. Throughout the year the current tends to maintain a general southward flow.

OREGON

Eureka

Pt. Reyes

San Francisco

Santa Cruz

Monterey

Pt. Sur

CALIFORNIA

NEVADA

Morro Bay

San Luis Obispo

Santa Barbara

Pt. Conception

Ventura

Los Angeles

San Miguel I.

Santa Rosa I.

Santa Cruz I.

Anacapa I.

San Pedro

Santa Barbara I.

San Nicolas Island

Santa Catalina Island

San Diego

San Clemente Island

ARIZONA

MEXICO

Pacific Ocean

If you examine a map of California, you will discover that the mainland coast south of Pt. Conception is sharply indented to the east. This identation is referred to as the Great California Bight. All eight California Channel Islands lie roughly in the lee of this bight. The water surrounding the Channel Islands is less immediately affected by the California Current than are waters to the north. South of Pt. Conception the current does not meet the coast of western North America again until just about 100 miles south of the Mexican border along the shores of Baja.

There is, however, a less dominant current named the Davidson Current which brings warm water north, particularly near the coast. As the northward and southward movements of great masses of water fluctuate, California waters often are affected by deep water upwellings of cold water that are pulled up or sucked up by the combination of winds and movement of the major currents. The upwellings are generally more prominent in Central and Northern California than in the south, and are also more common during spring and summer. These upwellings are nutrient rich, and play a major role in the life cycle of many species. They also have a significant impact upon water conditions, especially visibility.

Concerning Thermal Protection

From the Oregon border to the Mexican border, throughout the year, in California water divers NEED thermal protection. That means for dives of any depth and any duration, you need a wetsuit or a drysuit. Which of those systems you prefer depends primarily upon your build, your typical underwater activities, whether you dive on bad days as well as great ones, and, of course, your pocket book. If you decide on a wetsuit, in California almost all divers prefer one that is 1/4" thick. I can't tell you how many times I have heard someone say that in such and such a place they only need a 1/8" suit and for just a day or two in California water they are sure it will do. These divers may be well intentioned, but in California, divers require more thermal protection. A thin suit might enable one to survive, but diving is supposed to be fun, not an endurance contest.

El Niño

Every 2 to 10 years or so, our planet experiences a severe dislocation of the world's largest and most dominant weather patterns, a climatic phenomenon known as El Nino (pronounced el neenyo). The name, El Nino, is derived from the Spanish allusion to Christ, The Child, a reference to the dramatic increase in water temperature along the coast of South America which is usually noted near Christmas during the existence of the El Nino weather pattern.

As a result of the El Nino condition, weather in many regions of the world is dramatically altered. For reasons still unknown, weather systems that "normally" govern the tropical Pacific change and intensify. The direction and intensity of the equatorial trade winds and major ocean currents shift markedly. This change causes massive quantities of warm water to pile up along the western coast of South America. The warm water tends to move northward in a current that is commonly called the El Nino current.

Because so many weather factors are influenced by El Nino, the net effect in California can vary from severe droughts such as those experienced during the El Nino of 1976-77 to the intense winter storms such as those that inundated the state during the intense El Nino of 1982-83.

The most immediate effect as far as swimmers and divers are concerned is that water temperatures are often elevated by between 4° and 10° F, and can be increased as much as 14° F. That might not sound like a lot, but if you consider the fact that on a volume basis, water requires 3,600 times as much heat as does air to bring about a comparable temperature change, you can begin to realize how much energy the current contains. Consider, too, how important water temperature is to most marine life, and the picture becomes even more clear.

During the notorious El Nino of 1982-83, miles and miles of healthy kelp forests and the associated kelp communities were decimated by the rise in water temperature, and it took 2 years for many kelp beds to even begin to refurbish themselves. Conservationists at the Pt. Reyes Bird Observatory recorded unprecedented mortality rates among seabirds that breed on the nearby Farallon Islands.

On a worldwide scale, the elevated temperatures can have a devastating effect on many forms of plankton and all food chains that depend upon plankton as their base. At the same time, in California water many species of both migratory and benthic fishes were seen far north of their normal range. During the El Nino of 1982-83 wahoo, trigger fish, butterfly fish, and angel fish were common sights at some of the Channel Islands. Howard Hall, a well known underwater photographer, and Jon Hardy, a well known authority in the sport diving business, enjoyed the good fortune to film manta rays at Anacapa Island and Catalina Island, respectively. The northern range of mantas, like many of the just mentioned species, is normally considered to be the southern tip of the Baja Peninsula, over 1,000 miles to the south.

Water Visibility

Scuba divers and snorkelers use the term "water visibility" to describe the degree of water clarity. Water visibility, or "the vis," is expressed in terms of feet, indicating the distance a person wearing a diving mask can see underwater. Water visibility varies greatly depending upon when and where one enters California water. It is generally accurate to state that visibility varies from 5 to 10 feet off the beaches of the north coast to 150 feet at the southern Channel Islands. South of Point Conception water conditions are almost always considerably better than to the north. South of the point, visibility is often 5 to 15 feet greater, the water is 5 to 10° F warmer, and there is less wind and surface chop. However, there are few absolutes when it comes to west coast water. At times, water conditions in Northern California are superb, and when they are, you can safely bet that the snorkeling and diving are extraordinary.

The Channel Islands are a feature attraction for sport divers. Visibilty at the islands is almost always far better than off the beaches of the mainland. At the southern islands (San Clemente, Santa Catalina, Santa Barbara, and San Nicolas) visibility normally ranges from 50 to 100 feet, but 150 feet is certainly not unheard of. At the four northern islands (Anacapa, Santa Rosa, Santa Cruz, and San Miguel), visibility ranges from 20 to 80 feet on most days, but once again 150 foot days are not especially rare.

The Oceanic Provinces

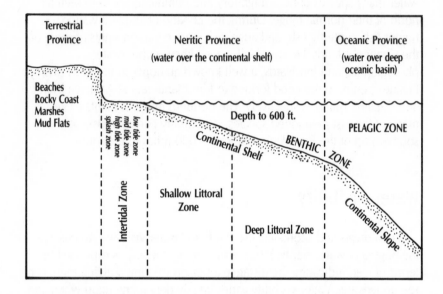

The sea floor from the coast of North America to a depth of about 600 feet is called the continental shelf. Along the majority of the California coast, the bottom slopes rather gently from the beaches to the edge of the shelf. While the slope is generally uniform in any given area, there are significant variations in the angle of the drop off in different sectors along the coast. This means that the width of the continental shelf varies considerably from one area to the next. South of the city of Monterey, for example, the shelf width is generally less than 5 miles, while north of San Francisco it extends to 20 miles in many places.

There is an abundance of life forms found between the beach and the edge of the continental shelf, both on the bottom and throughout the water column. In scientific discussions, the area along the bottom is referred to as the benthic zone. Almost every large marine plant and over 90% of the animal forms found in the world's oceans live in close association with the sea floor, and are, therefore, referred to as benthic organisms. The life forms that live primarily in the water column are called pelagic organisms. The pelagic zone is separated into two provinces, the neritic which includes water over the continental shelf, and the oceanic which includes the waters over deep oceanic basins.

Benthic animals range from the intertidal zone to deep oceanic trenches which can plummet to depths greater than 30,000 feet. Benthic plants, on the other hand, are restricted to a region where there is sufficient light to accommodate growth, a sector classified as the photic zone. The depth of the photic zone varies in accordance with those factors that affect light penetration through the water column. Thus, the photic zone reaches far deeper in tropical areas (where in some places it extends to almost 300 feet) than in the murkier, coastal waters of temperate areas. In California waters, benthic plants can be found to depths of about 150 feet. Beyond the photic zone is the aphotic zone, the region where the low level of sunlight inhibits the process of photosynthesis.

The benthic province is generally subdivided into 3 sectors, the intertidal, the shallow littoral (often called the sublittoral), and the deep littoral. The intertidal region is roughly defined as that portion of the coast which is exposed at low tide and covered at high tide. The life forms found within this region are of fundamental importance to many food chains.

Vertically arranged and to some extent segregated, distinct populations of plants and animals are commonly encountered in the intertidal zone in a configuration called vertical zonation. As the tides advance and recede over the intertidal zone, a number of factors such as temperature, light intensity, degree of predation, and the composition of the substrate vary greatly, creating conditions which lead to vertical zonation.

This pattern of zonation is continued to some degree in the shallow littoral zone which is basically defined as the sector between low tide and the area within the sea where plant and animal life become rather scarce. The shallow littoral zone extends from the mean low tide mark to a depth of approximately 150 to 200 feet, encompassing the region where sport divers do the vast majority of their diving. Most of the plants and animals discussed in this text inhabit the shallow littoral zone. Beyond 200 feet is the deep littoral zone, which is defined as that area between the deepest portion of the shallow littoral zone and the edge of the continental shelf, where the ocean plunges into the bathyl, abyssal, and hadal zones.

Red Tide

Red tide is the name given to a condition that occurs sporadically in waters off of the California coast. During a red tide the water takes on a dark reddish hue and water visibility is markedly reduced due to the presence of large quantities of a bacteria. In the northern part of the state the primary culprit is named Gonyaulax catenella, while Gonyaulax polyhedra is the dinoflagellate usually responsible for red tides in southern waters. The scientific community remains uncertain as to the triggering agent or agents that cause red tides.

In northern waters shellfish consume and accumulate large quantities of the bacteria Gonyaulax catenella while grazing. If the shellfish is subsequently eaten by humans, the result can be a serious, or even fatal, form of paralytic food poisoning. To prevent the problem, every year there is a quarantine on mussels in certain areas during portions of late spring, summer, and early fall. Public health officials have devised accurate tests to determine toxic levels. The bacteria is not absorbed through the skin, and there are no known health hazards to swimmers or divers who do not eat the bacteria laden mussels.

Gonyaulax polyhedra is bioluminescent, and fish leave bright trails of glowing light behind them as they swim through areas experiencing a red tide caused by that bacterial form.

Offshore red tides tend to occur in easily discernible patches, but in harbors and bays they tend to affect the entire area. Where water circulation is limited by the lay of the land, the red tide condition can persist for months. In some harbors and bays, extensive red tide conditions can lead to massive die-offs of a number of affected species. The bacteria die off normally, but their decomposition utilizes so much oxygen in the process that many other animals die from oxygen starvation as a result.

Geographical Divisions of California

Because of the length of the state and the north/south orientation, there is a much greater variety of conditions and life forms found in and along our coastal waters than are found in many other well known wilderness areas. Underwater California is not one homogenous site, instead being a large region filled with a wide variety of settings and a tremendous diversity of life forms. This great mixture of geophysical and biological phenomena contribute greatly to the allure of California's marine kingdom. The combination of stunning resources truly offers something for every recreational enthusiast.

When discussing California, it is a common practice to divide the mainland into 3 geographical sectors, Southern, Central, and Northern. With over 1,100 miles of coastline, the same division logically applies to the state's underwater world. Southern California is located between Point Conception and the Mexican Border. Central California is found between Point Conception to the south and San Francisco to the north, while Northern California is generally considered to be the area between San Francisco and the Oregon border.

The division of California by geographical boundaries coincides roughly with many, though certainly not all, of the differences in natural history found within these regions. While many of the plants and animals found in Southern California can be found in waters throughout the state, the waters of Southern California support an abundance of warm water life forms not normally found in waters further north. The flora and fauna in Central and Northern California are much more similar to each other as both regions lie to the north of Pt. Conception. There are, however, significant differences in the species encountered in the waters of Central California and Northern California as well as a major differences in the concentrations of specific populations.

The Habitats

As with the plants and animals found on land, marine plants and animals show distinct preferences for various habitats. As is so often noted by fishermen and divers, the habitats and bottom terrain vary considerably throughout California waters. Divers encounter dramatically different settings when they explore the kelp forests, the rocky reefs, open ocean seamounts, the sandy plains, deep water canyons, and the irridescent blue of the open sea. The changes in seascape are accompanied by significant differences in the plant and animal life found within. For example, a high percentage of the animals seen in the sand flats are rarely, if ever, observed in a kelp forest community. Likewise, the residents of a kelp community are seldom seen far out to sea. By gaining a basic understanding of these different habitats, you will immediately gain valuable insights into the adaptations of the plants and animals that live there.

It is, however, very important to understand that some animals often cross the boundary from one habitat to the next. Some species can be found in one habitat as juveniles and another as adults, or perhaps they feed in one locale and rest or breed in another. Still other species are frequently found in several habitats. But as a general rule, most animals prefer to dwell in one setting or another.

While scientists classify the various marine habitats in a variety of ways—according to the amount of light received, the effect of tidal flow, the depth, the location as it relates to the continental shelf—most people tend to describe the habitats in another way. We tend to refer to the different areas by their most conspicuous physical features. Thus the marine habitats in California are commonly referred to as: **the tide pools, the sandy beaches, the marshes, the mud flats, the kelp forests, the rocky reef, the sandy plains or deserts, and the open sea.** It is from this perspective—by habitat—that the plants and animals of California's marine kingdom will be discussed in this text.

Life at the Beach

Life at the Beach

From a biological perspective, the legendary beaches of California can be divided into three distinct habitats. They are the rocky beaches, the sandy shores, and the mud flats or marshes. These habitats are utilized as a home, resting place, feeding station and breeding grounds by a ' wonderful variety of marine animals. Collectively, the creatures that inhabit these biomes play a fundamental role in the overall ecology of both land and sea, serving as vital links in food chains involving birds, fishes, and many other life forms including human beings.

Animals that inhabit the various beach habitats display an amazing variety of adaptations that allow them to survive rapidly changing conditions. Many of these animals are able to manage for extended periods of time both on land and in water. And while on land, their immediate surroundings vary from extremely dry sand or rocks to damp sand and mud, to mixtures of sand and mud that can almost be described as liquid. Beach dwellers must also be able to tolerate radical temperature fluctuations, changes in weather and tidal conditions, and variations in salinity. Many are very motile in a vertical sense, meaning they can rapidly move up and down the beach as the tides move in and out. With the ever-changing conditions, the availability of food often changes rapidly as well. Large sources of food can be present one moment and washed away the next, presenting a problem that is not often experienced in other habitats. As a result, beach dwellers have adapted an array of methods to help maximize their feeding opportunities.

While few people explore the marshes on a regular basis, many of us spend a great deal of time walking along the rocky coasts and sand beaches, where we commonly observe a number of organisms. For that reason this chapter will focus primarily upon those animals who reside in the tide pools of the rocky outer coasts and those found in and on the sand. Many of the marsh animals that people commonly encounter are birds, and the most common bird species are covered in the chapter entitled "California Marine Birds."

It is of fundamental importance to be aware that many of the animals which are frequently seen in the tide pools (such as garibaldi, lobster, and abalone) and in the lower portions of sandy beaches (such as many species of crabs and sculpin) are also found in either the reef community or the sandy plains of the sea floor. These species are classic examples of animals that have adapted to a life style which allows them to survive and flourish while spending part of their lives in one habitat and another portion of their lives elsewhere. In fact, many species of crabs can survive on both land and in water. Considering that fact, I have had to make a decision about where in this guide to include those creatures who are frequently seen in several places. Rather than list all the information twice, I have chosen to briefly discuss two groupings of animals in the section on the beaches, which immediately follows. They are 1) some of the most prominent animals, and 2) those creatures that are only rarely, if ever, seen in other habitats. As examples, I have covered chitons in the section on tide pools, sand crabs in the pages devoted to life at the sandy beaches, garbaldi in the chapter entitled "The Reef Community," and California gulls in the section named "California Marine Birds."

California Tides

The tides in California are classified as mixed semidiurnal, meaning two high tides and two low tides occur each day, but the height of successive high and low tides is different. Tides are the result of the relative position and therefore, the gravitational pull of the earth, sun, and moon. Because those factors are well understood, the times and heights of various tides can be quite accurately predicted. That information is regularly printed in a number of readily available sources.

The Tide Pools

Along the rocky outer coasts of California, tide pools are inhabited by dozens of species that have successfully adapted to the rigorous conditions of life that exist in this narrow band where the sea collides with solid land. Many of these organisms live their entire lives in the tide pool habitat , but others spend only a portion of their lives in the tide pools or on the adjacent rocky shores.

Tide pools are located on rocky beaches where the tidal ebb and flow submerge the rocks, and then recede, leaving behind a basin containing enough water to sustain various forms of marine life. Meaningful tide pools vary in size from that of a small cup to small lakes.

No other place on earth experiences such drastic changes in living conditions on a daily basis as does the narrow coastal strip where the tide pools are found. Sometimes the tide pools are completely devoid of water; at other times they are flooded. Temperatures vary from extremely hot to below freezing. In some portions of the intertidal strip residents are frequently exposed to direct sunlight for extended periods. Often, wave after wave mercilessly pounds the tide pools and their inhabitants for days, weeks, or months on end. Biological factors such as predation and competition for living space also play important roles in the dynamics of tide pool life. In short, living conditions are both extremely varied and extremely harsh. And yet, amazing as it sounds, on a global scale, this strip is perhaps the earth's most densely populated region.

In order to survive in such a demanding environment, organisms must be 1) able to firmly attach themselves to the rocks, 2) be encased in a strong shell, or 3) be especially adept at hiding in between rocks, or in small crevices in order to protect themselves, and 4) be able to prevent desiccation, which is the fatal evaporation of water from body tissues. Desiccation is the greatest killer of all tide pool organisms.

The basic needs of the plants and animals that live in the tide pools are identical to those of life forms that exist elsewhere. All need food, oxygen, and water, and some require shelter. That might not seem like too much to ask, but meeting those needs presents a daily, if not hourly, challenge to tide pool organisms. A multitude of astonishing adaptations which are fundamental to the dynamics of tide pool life enable the various organisms to survive. Some of these adaptations include the ability of many species to control water loss during low tide, the use of a shell for protection and as a means of helping control water loss, and comparatively large surface areas on the gills of many tide pool species which enables these animals to maximize the amount of oxygen that is available. Possessing large gills is especially valuable to tide pool organisms because the oxygen content in the pools is comparatively low especially during low tide. A number of physiological adaptations which afford for greater efficiency in the use of oxygen also prove to be highly beneficial to tide pool organisms.

The intertidal region of the rocky beaches—that area which is immediately impacted by ebb and flow of the tides—is generally divided into four zones. They are **1) the splash zone, 2) the high tide zone, 3) the middle tide zone,** and **4) the low tide zone.** Exactly where the boundaries are drawn, and whether or not those are all the zones remain points of scientific debate. Certainly the boundaries for these zones are artificial in the sense that they are frequently and consistently crossed by their residents. In addition, a close examination of a rocky beach will likely reveal that a species that obviously dominates one zone can be found in other zones as well. However, in the final analysis it is also obvious that various plants and animals do show strong preference for the conditions found in the various zones.

As you might expect, the environmental conditions found within each zone differ considerably, and this leads to a definite striation or layering of life forms found within the intertidal strip. This layering is known as vertical zonation, and is a phenomenon that continues to some degree beyond the tide pools into the sublittoral zone. Vertical zonation, common in all coastal habitats, dictates that different species of animals will choose to live in different places relative to the height of the tides. The phenomenon occurs because environmental factors such as temperature, light intensity, and oxygen content vary as tides advance and recede.

The splash zone, as the name implies, lies above the high water mark and is affected only by the splash or spray of waves. This area is only rarely inundated by extremely high tides and conditions produced by severe storms. Yet a surprising number of limpets, copepods, isopods, periwinkles, crabs, and barnacles survive quite readily in the splash zone. Several species of green and blue-green algae also thrive in the splash zone environment. These algaes often grow on tarlike bands or patches marking the upper bounds of the splash zone. The algaes are remarkably tolerant of dramatic changes in temperature, and are especially adept at resisting desiccation, a result of their ability to form a gelatinous mass that traps and stores significant amounts of water.

A few species of snails, limpets, and some crustaceans graze on the algaes found in the splash zone. The snails and limpets are especially able to tolerate temperature changes, and both can seal the edges of their shells to the rocks in order to retain moisture. If you look closely at a snail or limpet in the splash zone, it is likely that you will find these species to be tightly clamped down on the rocks. This is especially true during low tide, a period during which these animals have been and will be without the benefit of spray.

Just below the level where the snails and limpets occur, you are likely to discover concentrations of small acorn barnacles in the genus *Balanus* and *Chthamalus*. These barnacles are filter feeders, but amazingly enough are able to survive while feeding only a few hours a month, during the highest tides. During those tides the barnacles extend their feather-like feet which are used as feeding appendages from their volcano-like shells as they seek tiny forms of plankton in the water around them. Between the high tides, these remarkable animals seal themselves inside their shells through the use of a calcerous plate which blocks the entrance.

The high tide zone is best described as the area between the mid-tide level and the region covered by most high tides. The creatures that live in the high tide zone must be equally adept at surviving on both land and sea, as this zone is awash twice each day. This zone is populated by far more species than the splash zone. In fact, the further you move away from the splash zone the more life forms you are likely to discover.

A variety of crabs, snails, limpets, mussels, and plants also inhabit this region. While many of these species fill a necessary niche in nature's scheme, most hobbyists can not readily differentiate between various species of crabs, snails, and copepods etc.—and most of us find it unnecessary to even try. Many of these organisms are also seen both in the splash zone and in deeper waters. As far as the variety of animal species is concerned, the most significant difference between the high tide zone and the drier environment of the splash zone is that there are more species of arthropods and mollusks found in the high tide zone.

The middle or mid tide zone is the region between mean sea level and the mean low water line reached during low tides. This area abounds with mussel beds, sea stars, brittle stars, many species of fish, anemones, barnacles, nudibranchs, crabs, snails, worms, adult black abalone, some adult green abalone, many juvenile abalone, octopus, lobster, sea hares, and more.

Speaking in strict scientific terms, the most dominant species found in the middle tide zone are mussels (in the genus *Mytilus*), barnacles (in the genus *Balanus*), and a variety of chitons. These species are prominent visually as well. If you explore a rocky beach or examine a pier piling when the tide is out, odds are good that you will be able to find a number of mussels, barnacles, and chitons. These animals have rounded, low profiles and are able to anchor themselves securely to the substrate to minimize the pounding from the ever present surf.

Several species of algae are dominant as well. Like the animals that are able to survive in tide pool areas that are constantly exposed to breaking waves, these plants are strongly anchored to the bottom. The algaes utlize strong holdfasts, and flexible, yet sturdy, stipes to secure themselves.

Mussels, barnacles, chitons, various forms of algaes, and several other bottom dwelling species must compete for both space and food in the mid tide zone. While space is at a premium, nature has devised away to constanly provide some unused sea front property. Barnacles and mussels are constantly preyed upon by starfish (sea stars), limpets, and carnivorous snails. Not all limpets are carnivores, but even those that are not cause considerable damage to barnacle populations as the limpets often dislodge the barnacles while grazing on algae. Driftwood, heavy surf, and seasonal variances cause some die off in various algaes.

If left undisturbed, mussel populations will usually out compete barnacles. However, starfish rarely allow conditions in mussel beds to remain undisturbed for extended periods of time. This is especially true in the lower portions of the mid tide zone and in the low tide zone, as starfish are especially vulnerable to desiccation. Snails, too, prey upon mussels in significant amounts. The mussels combat their predators in two ways. First, the mussels attempt to become so numerous that their predators can not possibly devour them all. Second, as the mussels become larger their shells become thicker and harder preventing drilling snails from penetrating the shells and getting to softer body tissues. It is important to realize that the process is ongoing and there is never a clear cut "winner" or "loser." The truth in that statement can clearly be understood by realizing that as patches of mussels are cleared away by pounding waves or by predation, that space is quickly taken over by algae and barnacles. Given time, however, the mussels regain the upper hand as the process of biological succession continues.

Several species of anemones such as the aggregate or matted anemones *(Anthopleura elegantissima)* inhabit the tide pool areas found in the middle tide zone. The upper range of anemones is generally determined by their ability to prevent both fatal desiccation and greatly elevated body temperature. Aggregate anemones have adapted well, and studies have proven that these animals can withstand internal body temperatures that are up to 55° F above the surrounding air temperature. To prevent the fatal loss of water from body tissues during low tide, these anemones often retract their tentacles and cover themselves with light colored rocks and shells which tend to reflect, rather than absorb, heat.

The low tide zone is exposed to air only during the lowest tides of the year. Diversity of species, not domination by only a handful, is the key characteristic of the low tide zone. Here you'll discover an array of animal species including many fishes, sponges, morays, nudibranchs, sea hares, anemones, chitons, sea cucumbers, limpets, shrimp, urchins, sea stars, brittle stars, hydroids, clams, crabs, and a variety of snails. Eel grass, surf grass, sargassum, and red and brown seaweeds are the dominant plants.

The density of populations of echinoderms—sea stars, sea cucumbers, brittle stars, and sea urchins—increases markedly in the low tide zone. These animals are highly susceptible to the threat of desiccation in higher tidal zones, but thrive in the low tide zone as the number of potential food sources also increases.

Observing Tide Pool Life

Of all the places where an enthusiastic naturalist can observe marine life, no habitat affords a better opportunity than the tidepools. Animal behavior and interaction among various species can be witnessed in great detail over extended periods of time almost any day or night of the year. Access to rocky beaches and tide pool communities is good in many locations all along the California coast. And as opposed to Scuba diving or snorkeling, your time as an observer is not limited by your supply of air. A series of leisurely spent mornings, afternoons, and evenings in various tide pool communities during different tidal conditions will go a long way in providing fascinating insights into one of the earth's most incredible biomes.

Learning about the plants and animals that inhabit tide pools and rocky beaches will also be a lot of fun, but please be careful only to observe and not to disturb and disrupt. A significant amount of harm can be done through unthinking acts if people are not careful. For example, the removal of rocks from the retracted tentacles of an anemone can provide people with a moment of pleasure, but for the anemone the threat of dessication can be life threatening. Be a thinking naturalist! Observe and enjoy, and try not to intercede.

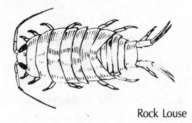

Rock Louse

Phylum: Arthropoda
Rock Louse

The rock louse *(Ligia occidentalis)* is sometimes called the common rock isopod. This creature is a crustacean, as are all 4,000 species of isopods. The term isopod, meaning "similar feet," refers to the similar appearance of 7 pairs of thoracic appendages.

Usually seen in great numbers, rock louse are especially active at night, and are often seen darting about on the rocks in the upper reaches of the splash zone, well above the high tide line. Rock louse can generally be described as flattened from top to bottom, appearing quite similar to pill bugs, or "roly polys" that are found in gardens. These arthropods attain a length of 1.5 inches, and although their coloration varies, they usually blend in well with the surrounding rocks. Most specimens are darkly colored, almost black. Certainly that is the case with the common rock isopod. Rock louse are scavengers, feeding primarily on organic decay.

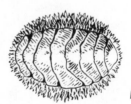

Mossy Chiton

Phylum: Mollusca
Chitons

Chitons look like small oval shaped mounds that are covered with a sectioned plate of armour. Their unusual shape has led to a number of other names including "sea cradles" and "coat of mail" shells. Though they are mollusks, they are the only mollusks that have jointed shells. The flexibility afforded by the 8 calcerous plates of the upper shell allows chitons to bend so they can readily fit into uneven depressions in the rocks. The plates are surrounded by a girdle which enables the chiton to seal tightly to the rocks in an effort to prevent dessication, the rapid, fatal evaporation of water from body tissues.

Most chitons hide under the rocks or in depressions in the surface of rocks during daylight, but you can easily find them during the day or at night on almost any rocky coast. These mollusks grab onto the substrate with their broad, strong foot. Herbivores, chitons come out to feed on various algaes at night.

Three species of chitons are commonly seen in California tide pools. They are the mossy chiton *(Mopalia mucosa)*, the troglodyte chiton *(Nuttallina californica)*, and the conspicuous chiton *(Stenoplax conspicua)*. Mossy chitons grow to a length of about 2.5 inches, are dark brown to green, and their fleshy girdle is covered with stiff hairs. Seaweeds often attach to mossy chitons. Mossy chitons are found throughout California, and have a tendency to be slightly larger in the northern part of the state. Most troglodytes are only about 1.5 inches long. Usually found in either the high tide or middle tide zone, these chitons tend to permanently inhabit shallow depressions on the tops of large rocks. Troglodyte chitons cut a pit, or "sear," into a rock to match their own shape. Troglodyte chitons are extremely territorial, and during a life that can span up to 20 years, they almost never leave the pit, except to graze during periods of high tide. After feeding, they return to the same pit. The coloration of troglodyte chitons varies from light yellow to dark brown.

Conspicuous chitons are the largest of the California chitons, reaching a length of 6 inches. They have gray-green to brown bodies with some pink.

Bay Mussel

Mussels

Mussels are among the most dominant and most conspicuous of the members of the upper middle tide community on rocky outer coasts. California sea mussels *(Mytilus californianus)* live in densely populated beds in this portion of the intertidal zone.

In their adult stage, mussels attach to the substrate by using strong, elastic, hair-like structures called byssal threads which are produced by a gland in the foot. Although mussels are capable of movement, once they attach they rarely change positions.

Each mussel must compete with other mussels, barnacles, and algae for both space and food. If left alone, mussels will outcompete barnacles and algae, but they are rarely left undisturbed for long, as they are heavily preyed upon by several species of sea stars. Mussels are filter feeders and capture their food by straining the water for tiny plants and animals.

The bay mussel, *Mytilus edulis,* is a dark blue to black mussel that is often seen on wharf pilings. These mussels also inhabit back bays, estuaries, and waters from the intertidal zone to a depth of 120 feet.

Goose-neck barnacles are often found in mussel beds, as are several types of crabs, clams, worms, shrimp, and hydroids. These animals are highly dependent upon the existence of mussel beds for their own survival. Goose-neck barnacles use their foot as a net to capture food. When the barnacle is submerged, the foot is extended to strain the water for prey.

Flat-bottomed Periwinkle

Periwinkles

Two species of periwinkles are often seen high up in the splash zone, well above the level of the highest tides. They are the flat-bottomed periwinkle *(Littorina planaxis)* and the checkered periwinkle *(Littorina scutulata).* In fact, flat-bottomed periwinkles live most of their lives out of water, occupying a higher vertical position on shore than any other California marine mollusk. Periwinkles are snails, members of the class of gastropods, as are abalone and limpets. A link between the creatures of the land and sea, periwinkles must keep their gills wet in order to survive, but they can not remain indefinitely submerged because they will drown.

Flat-bottomed periwinkles are small, growing no longer than ¾ inch. The shell, which is gray except for a faint white line along the lower border of the aperture, contains no more than 4 whorls, or spirals. The distinguishing feature of these snails is the greatly flattened central axis on the lower surface of the shell. Flat-bottomed periwinkles are rarely solitary.

Checkered periwinkles can be identified by the light colored spots on the shell and the absence of the white line found on flat-bottomed periwinkles.

Phylum: Crustacea
Crabs

Striped Shore Crab

Pea Crab

While many species of crabs inhabit rocky beaches, at least four species warrant special mention. They are the flat porcelain crab *(Petrolisthes cinctipes)*, the blue-clawed hermit crab *(Pagurus samuelis)*, the pea crab *(Pinnixa littoralis)*, and the striped shore crab *(Pachygrapsus crassipes)*. Scavengers, flat porcelain crabs are quite common in the middle and low tide zone. They are often found clinging to the underside of rocks. If you turn the rocks over, the crabs will quickly run for cover. These crabs can be distinguished by their long antennae and their flattened bodies.

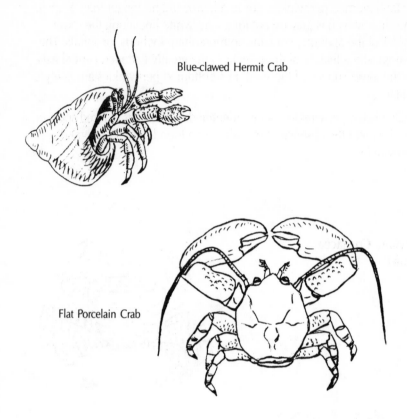

Blue-clawed Hermit Crab

Flat Porcelain Crab

The blue-clawed hermit crab is the most common of the hermit crabs found in Southern California tide pools. These crabs appear to prefer to live in the abandoned shell of black turbans. Blue-clawed hermit crabs are found from Alaska to Baja.

Pea crabs are commonly found living inside the gill chambers of California mussels, where they constantly pick at the gills removing particles of food causing considerable harm to their hosts. The crabs, one to a mussel, usually leave the mussel only to breed. Pea crabs are rather small, most being less than ½ inch in diameter.

Striped shore crabs are the most common shore crab in Southern California. They can be distinguished by their diameter which is usually close to 2 inches, their large, dark red to purple claws, and their abundance. Although these crabs also reside in the mud and sand, they are most common along rocky outer coasts.

The Sandy Beaches

Sandy beaches exist where waves are gentle enough to allow sand, grains of quartz, and animal skeletons to accumulate, yet still powerful and constant enough to wash away silt, mud, and clay. Sandy beaches are comparatively unstable. They tend to shift, change character, and conform to the conditions imposed by tides, waves, and currents. Few plants can exist in such an environment, but some small animals do manage quite well in these demanding surroundings. Nevertheless, when compared to the varied and prolific populations of the tide pools and rocky beaches, the sandy beaches are comparatively desolate. The lack of a stable substrate prevents those species that need to securely anchor themselves from settling on the sand.

Many of the animals that do inhabit the sandy beaches also spend a significant amount of their lives completely submerged. For example, some "sand crabs" might at times be seen high and dry on a sandy beach, and a short time later the same animals might be observed in more than 50 feet of water. Most of the larger sand dwellers are excellent burrowers, while many smaller animals can actually move among grains of sand without much displacement. A high percentage of sand dwellers are detritus feeders—animals that feed on decayed bits of organic matter that is washed toward shore.

Phylum: Arthropoda
Amphipods

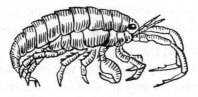

Large Beach Hopper

Tiny amphipods, commonly called beach hoppers and scuds, literally cover the upper portions of many California sandy beaches at night. Also called sand fleas, amphipods are actually crustaceans, not insects. Marine amphipods occur in habitats that vary from the highest portion of the high tide zone to abyssal depths of the sea floor in oceans all over the world. Operating on a 24 hour cycle, or circidian rhythm, coastal dwelling amphipods tend to burrow into the sand to hide during the day. Beach hoppers are much more active at night than they are during daylight hours. These amphipods derived their common name from their uncommon bounding method of moving on the sand.

Several species of amphipods inhabit California beaches. In Southern California the most common is the large beach hopper *(Orchestoidea corniculata)*. These amphipods tend to reside in areas of coarse, damp sand. Large beach hoppers attain lengths of up to one inch. An even larger species *Orchestoidae californiana* is quite common north of Los Angeles. Beach hoppers generally prefer to hide under debris and drift kelp, or to burrow in the sand during daylight hours. Often you can find beach hoppers by overturning debris and kelp on the beach.

Common Sand Crab

Crabs

Several species of sand crabs or mole crabs live on beach front property in California. These crabs burrow into the sand near the tide line. They usually bury their entire body except for their eyes and antennae. When buried, they point the "V" of their antennae away from the water. The antennae are used to trap food as the waves recede. Sand crabs feed on pieces of sea weed. Interestingly, sand crabs always move backwards, whether they are swimming, digging, or walking.

The most common sand crab is appropriately named the common sand crab *(Emerita analoga)*. However, this species has a rather uncommon appearance. The legs are modified for digging, the antennae are quite long, and the body carapace is elongate, and convex. Males are smaller than females, which attain a length of just over 1 inch. They are a pale blue to off white. Common sand crabs feed by trapping food with their long, feathery antennae. The crabs burrow into the sand tail first, and face toward the water. As waves break over the crabs, the animals extend their antennae and trap plankton and bacteria which is then transported to their mouth.

Common sand crabs are often seen scurrying along the beach, remaining close to the water's edge as the tide moves up and down. These crabs are a favorite bait of surf fishermen.

Pismo Clam

Phylum: Mollusca
Pismo Clams

Pismo clams *(Tivela stultorum)* are mollusks that inhabit some sandy beaches between Half Moon Bay in Central California and Magdalena Bay about half way down the Pacific side of Baja. These clams derive their name from Pismo Beach, where they were once quite common. Pismos spend their lives burrowed about 10 inches deep in the sand, feeding and breathing through a muscular siphon tube. There are a pair of holes at the top of the tube. One is incurrent, used to draw in water that is filled with food and oxygen. The other is excurrent, used to eliminate wastes.

The exterior of the shell of Pismo clams is tan to dark brown, and is often covered with medium brown stripes. A delicious but overly hunted food source, pismos reach a length of about 5 inches in 4 to 7 years. Usually where you find one pismo clam, you can find several. A good way to locate pismos is just to walk along the beach looking for water to suddenly "jet" out of the siphon. If you step close to the clam, your weight will cause the water to shoot out of the siphon. And then you can dig the clam out of the sand as long as they are in season and you possess a fishing license as required by state law. Scuba divers, too, can take pismo clams as long as they comply with the regulations established by the California Department of Fish and Game. Underwater, the clams are best spotted by looking for their twin siphons, or the sipon holes, which protrude above the sand.

Other common California clams include the wavy cockle *(Chione unidatella)*, the little-neck clam *(Protothaca staminea)*, and the gaper clam *(Tresus nuttalii)*.

The Mud Flats and Marshes

Survival in the mud flats and marshes is often highly dependent upon an animal's ability to tolerate rapid changes in salt content in the surrounding water. An adaptation that allows some bivalve mollusks to survive is their ability to close their shells trapping water that has a sufficient amount of salts. These bivalves stop feeding when the salt content becomes too low, and they simply remain isolated within their shells until conditions become more favorable.

Some tunicates and anemones are capable of tolerating extreme variations of internal salt content, and are referred to as osmotic conformers. They are soft bodied, somewhat porous, and unable to control the flow of water through their own tissues, but they have greater levels of tolerance to variations in salt content than most animals.

Plants and animals that reside in the mud experience a similar problem, as the supply of oxygen is usually quite low in both the mud as well as the water that is in the mud. Many of these animals, such as several species of piddock clams, solve that problem by relying upon tubular siphons that reach the surface and fresh air, or simply reach up into water that has a higher oxygen content.

The creatures of the mud are usually much more active when the tide is in. When the tide is rising, the marshes and mud flats are often visited by shore crabs, shrimp, and fish that come to feed or spawn. Their stay is only temporary, as they depart with receding waters.

Birds such as long-billed curlews, sandpipers, oyster catchers, egrets, and herons feed on the marine creatures of the mud flats and marshes. Many birds that frequent the marshes have long toes for added balance, and long beaks with which they can probe into the mud for food. And land dwellers such as raccoons and foxes also feed in the marshes, especially during periods of low tide.

Life in the Kelp Forests

Life in
the Kelp Forests

A major attraction, and perhaps the most unique feature of California's underwater world, is the presence of magnificent kelp forests. The forests are so named because kelp plants grow in large patches· which can cover up to several square miles. These aggregations of kelp are referred to as beds or forests.

During ideal diving days, few places on earth appear more inviting than a kelp bed. On the surface, the Pacific is flat calm, the water is warm, and there is hardly a cloud in the sky. Underwater, shimmering rays of sunlight dance through the towering forests of giant kelp as waves pass gently overhead. The golden hues of the kelp fronds stand out against a blue-green background as strands of bright green eel grass flow with the surge along the rocky bottom. At a depth of 20 to 30 feet below the surface canopy divers become aware of the rhythmic sway of the entire forest as it keeps time with the ocean's movement. As far as one can see, the entire forest moves gently back and forth with the surge, the plants and animals within moving in perfect synchronization with the ocean's ebb and flow.

But on rough and stormy days, the shadow filled forests often look dark and ominous. When the sea is churning, the towering plants constantly tug at their holdfasts, often pulling free from the bottom and becoming entangled with other plants, which in turn entangle neighboring plants, creating a destructive domino effect. In conditions like these, the surge along the bottom can easily push and pull a diver 10 to 15 feet with each passing swell.

When conditions are good, the opportunity to observe a diversity of marine creatures in a beautiful wilderness setting is one that is only rarely, if ever, surpassed. Bright orange garibaldi, schools of silver colored Pacific jack mackerel flashing in the sunlight, and curious sheephead often greet divers as soon as they make a splash. Migratory fishes such as yellowtail, barracuda, black seabass, and white seabass occasionally visit the kelp to feed. More than 60 species of rockfish hover over the bottom, while cabezon, sculpin, gobies, convictfish, and many other colorful bottom dwelling fishes rest on the rocky substrate below.

Moray eels, California spiny lobsters, abalone, and a host of colorful invertebrates ranging from anemones to starfish to brilliant sea fans and purple coral await in the recesses of the reef. These are common sights, and you just never know when you'll be lucky enough to swim with a harbor seal, a herd of sea lions, a school of bonito, or look up after photographing a tiny nudibranch to see a gray whale swimming overhead.

Common Kelp Species

There are 21 species of kelp found along California's Pacific coast. Of these, several species are more prominent as far as recreational enthusiasts are concerned. Among them are: giant kelp (popularly called Macrocystis after its scientific name *Macrocystis pyrifera*), feather-boa kelp (*Egregia laevigata*), elk kelp (*Pelagophycus porra*), 3 species known as palm kelp (*Eisenia arborea, Poftelsia palmaeformis*, and *Pterygophora californica*), laminaria (*Laminaria dentigera*), strap kelp (*Eigregaria menziesii*), bull kelp (*Nereocystis leutkeana*), and argarum (*Argarum fimbriatum*). Though in some instances these species can occur in the same areas, there are significant differences in concentration in various geographical and oceanic zones.

Giant Kelp is the most dominant species in Southern California. Giant kelp grows from a rocky bottom that is usually between 25 and 130 feet deep. (See "Giant Kelp" page 62.)

Feather-boa kelp, sometimes called ribbon kelp, lines the intertidal region to a depth of about 50 feet. The species named feather-boa kelp attains a height of only 10 to 12 feet. Appropriately named, this kelp bears strong visual resemblance to long feathery ribbons, lacking both branches and long blades. Feather-boa kelp inhabits waters between Pt. Conception and Ensenada.

Elk kelp inhabits the waters south of Pt. Conception between depths of 10 to 90 feet. Elk kelp is characterized by its very long stipe which ends in a single, grapefruit-sized spherical float from which two rows of long flat fronds branch out. Elk kelp is often observed along the outer edges of giant kelp forests, where it can receive sufficient sunlight and is free from the domination of the larger Macrocystis plants.

The **palm kelp** scientifically known as *Eisenia arborea* can be found from intertidal depths to 120 feet from Monterey to Baja and can be distinguished by the serrated edges of the blades, and the stipe which forks just below the blades.

Palm kelp is also the common name given to a species that is scientifically known as *Poftelsia palmaeformis*. This kelp is a short plant that appears intertidally in Central California.

Also called **palm kelp**, the species that is scientifically named *Pterygophora californica* appears along the entire coast of California. The smooth blades of Pterygophora spread out from the upper half of a single stipe. This kelp grows from intertidal depths to approximately 120 feet.

Laminaria can be observed along the entire length of the coast, though it is most common in Central California. The stipe grows only to a height of about 1 foot before spreading laterally into a series of 3" to 6" wide blades that can reach a length of 6 to 8 feet.

Between Point Conception and San Francisco, the forests may consist solely of bull kelp, of giant kelp, of sargassum weed, of strap kelp, or a combination of any two, three, or all four. Present from Oregon to San Miguel Island, **bull kelp** is the most dominant species north of San Francisco. This plant can be identified by the streamer-like blades that are fastened to a tennis ball sized float, which in turn is attached to a rope-like stipe.

Strap kelp is found inshore north of Point Conception and can be distinguished by its long stipes bordered with short, thin blades.

Argarum is characterized by its wrinkled looking, hole filled, single blade that reaches a length of up to 8 feet and has a width of 2 to 3 feet. The blade appears perforated, being laced with a number of holes.

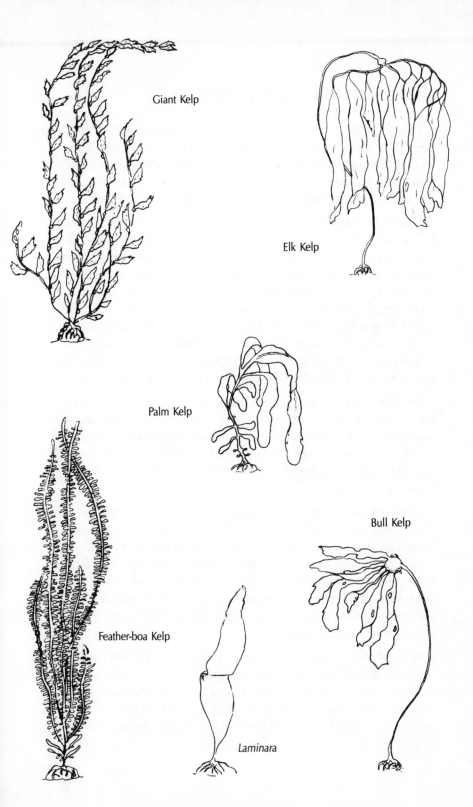

Giant Kelp

Elk Kelp

Palm Kelp

Bull Kelp

Feather-boa Kelp

Laminara

Giant Kelp: *Macrocystis Pyrifera*

Though there are often a number of species of kelp found in most kelp forests, the mainstay of the ecosystem of the largest forests which are found in Southern California is giant kelp, *Macrocystis pyrifera*. Giant kelp forests inhabit waters ranging from Santa Cruz, California to Turtle Bay, about halfway down the Pacific side of Mexico's Baja Peninsula.

Giant kelp is a brown seaweed, a type of algae, that has descended from plants which survived the ice ages. Seaweeds are not as complex as flowering plants, lacking special tissues that carry water and food from one part of the plant to another. However, in many forms seaweeds have demonstrated a remarkable ability to survive in rather demanding conditions. Giant kelp is a classic example. Kelp forests thrive in areas where there is too much water motion and currents for many other plants to survive. However, this water flow provides a continuously renewed supply of nutrients, such as nitrogen, potassium, and sodium, which are absorbed by the kelp plants and are vital to their survival.

The largest plant in the marine environment, giant kelp is also the fastest growing, capable of increasing in length by 2 feet a day under ideal conditions. In only 6 to 8 months of good conditions, a kelp plant can reach the surface from a starting depth of 100 feet. Healthy beds of Macrocystis are found only in regions where water temperature stays between 50°F and 68°F.

Giant kelp rarely grows from deeper than 130 feet, yet individual plants often reach lengths of up to 200 feet as the golden fronds grow straight up from the bottom to the surface where they stretch out horizontally forming a floating surface canopy. In a healthy kelp forest during summer and fall, the canopy can be up to 10 feet thick. The suspended overhang blocks a portion of the sun's light helping to create a stunning, cathedral-like effect for divers who easily have room to manuever below the canopy.

With very few exceptions, kelp beds are found in areas with a rocky substrate. Mature giant kelp plants are comprised of a holdfast and a number of buoyant fronds. The frond, that part of the plant above the holdfast, is comprised of a stem-like stipe, a number of leaf-like appendages called blades, and gas bladders. The blades attach to the stipe with connective tissues called pedicels.

The plants have no true root system, but instead depend upon a system of short, thin, sturdy structures called haptera, which look like oversized pieces of spaghetti. Collectively, the numerous haptera form a holdfast which attaches the plant to the bottom. The haptera do not penetrate the substrate as is the case with roots in true flowering plants. The holdfasts are well designed both for gripping the substrate and for resisting the constant shock and pull of wave action, surge, and current. However, severe winter storms, the biggest natural threat to kelp forests, can rip the holdfasts loose from the bottom. Unlike terrestial plants which take in most of their nourishment from their roots, kelp absorbs nutrients from the water through all of the plant's surfaces.

The haptera are generally incapable of attaching to sand, mud, or even silt covered rocks. This fact helps explain the demise of a great many kelp beds which were located near the sewage outfalls of major metropolitan areas during the 1950's, 60's, and early 70's. The sewage created a layer of silt on the bottom, preventing new holdfasts from attaching to the bottom and destroying miles of healthy kelp communities. In addition, high concentrations of phosphates and other chemicals in the sewage killed many existing plants. On the whole, Californians have become much more aware of the significant impact of our sewage, and most towns have taken responsible actions to prevent recurrence of similar problems.

Out of water kelp is quite heavy, and large entangled clumps often litter the beaches, especially after heavy winter storms. But in the sea, the fronds float upward as gas filled bladders called pnuematocysts buoy the plants. The pnuematocysts are found between the stipe and blades. Their buoyancy allows the fronds to reach the surface where the plants receive sunlight, an ingredient vital to photosynthesis, the process by which plants convert sunlight to energy. Unlike leaves on terrestrial plants, kelp blades have no "top" or "bottom" side. This feature enables the blades to conduct photosynthesis on both sides rather than only on the top. The end result is that kelp can grow very rapidly even though the blades are constantly being flipped over by water action. In fact, the entire frond takes part in the process of photosynthesis.

As with all species of kelp, giant kelp reproduces through a procedure known as alternation of generations in which a sexually reproducing generation alternates with an asexually reproducing generation. Thus, the complete reproductive cycle consists of two generations of plants.

The Kelp Community Ecosystem

Describing kelp beds as undersea forests is a valid analogy in many ways. It is not merely the foliage alone that divers find so alluring, for the beds provide living quarters for an estimated 800 species of marine life. In a typical kelp forest, there are so many marine organisms using the ecosystem for food, protection, and a substrate for attachment that it is difficult to calculate the sheer numbers. One fully mature kelp plant alone will support more than one million organisms, many of which are microscopic in size.

It was Charles Darwin who first noted the ecological importance of kelp forests when in 1834 he proclaimed: "The number of living creatures of all orders, whose existence intimately depends on the kelp is wonderful. A great volume might be written, describing the inhabitants of one of these beds of seaweed . . . I can only compare these great aquatic forests . . . with terrestrial ones in the intertropical regions. Yet, if in any country a forest was destroyed, I do not believe nearly so many species of animals would perish as would here, from the destruction of kelp." The last sentence of Darwin's statement is perhaps subject to question due to the discovery of so many previously undocumented insect and bird species in tropical rain forests. Nevertheless, the biological importance of kelp beds is well established.

It has been shown that over 178 species live in the holdfasts alone. This list includes creatures such as crabs, nudibranchs, brittle stars, isopods, and worms. Scientists working out of the University of Southern California Marine Science Center at Catalina Island have documented more than 100 species of motile invertebrates which are commonly found in and around the fronds. These species include a variety of shrimp, mysids, isopods, copepods, amphipods, and kelp snails. While most of the small invertebrates that inhabit the kelp community go unnoticed by divers, their presence serves to attract many of the 125 fish species commonly seen in kelp communities which feed upon these tiny organisms.

A snail commonly called Norris' top shell (*Norrisia norrisi*) is a beautiful species that is often observed on giant kelp plants. Its shell is a handsome brownish red, while the foot is deep red to orange. Although occasionally seen on the rocky substrate, these snails spend the majority of their lives grazing upon the kelp plant. Starting on the holdfast, the

California
Beaches

1. Rocky cliffs — typical of the Central and Northern Coast.

2. Coastline of world famous La Jolla.

Wildflowers flourish at many uninhabited beaches during spring and summer.

5. Marbled godwits feed along the shore.

4. Flock of cormorants at Carmel.

6. Dying red crabs awash at La Jolla Cove.

7. Juvenile garibaldi inhabit many Southern California tide pools.

Chiton exposed at low tide.

9. Solitary green anemone.

). Hermit crab.

11. Kellet's whelk.

2. Tide pools are created during receding tides when water is trapped in rocky basins.

California
Kelp
Forests

13. Surface view of giant kelp canopy.

14. Gas-filled pneumatocysts buoy kelp plants toward sunlight.

15. Holdfasts anchor kelp plants to rocky bottom

16. A school of jack mackerel cruise the kelp.

7. Sunlight shimmering through a forest of giant kelp creates a dramatic scene for snorkelers and scuba divers.

19. Underwater view of sunlit kelp canopy.

18. Kelp snails are commonly found on giant kelp plants.

20. Colonial hydroids on a blade of giant kelp.

21. California sea lions glide through a Channel Island kelp forest.

2. Black surfperch.

3. Giant kelp fish camouflaged in kelp.

24. Urchins readily feed on holdfasts when competition for food intensifies.

5. Protected by state law, garibaldi can be observed in many Southern California kelp communities.

26. Feather boa kelp (also ribbon kelp).

27. Palm kelp (*Eisenia arborea*).

28. Bat rays explore kelp forests looking for food on rocky reefs below.

snails slowly work their way to the end of a blade. At that point, they fall off the plant, crash into the bottom, locate a nearby holdfast and begin the process over again. Kelp snails make excellent photographic subjects with the bright red foot contrasted against the golden fronds. Kelp snails are often covered by a number of slipper shells. Slipper shells are mollusks that attach to the kelp snail's shell. The color of slipper shells is highly variable. Slipper shells are filter feeders, sometimes called suspension feeders, and are not parasites.

Of fundamental importance to the overall health of a kelp ecosystem is the fact that while individual kelp plants often live for several years, the blades have a life span of only a few months, after which they fall off and decay. The short lifespan of the blades prevents colonies of encrusting animals such as bryozoans and hydroids from overweighting the kelp frond, which would cause the plant to sink and thus prevent it from attaining vital sunlight at the surface. Equally as important, the constant natural process of blades and fronds dying and decaying, known as sloughing, provides the primary source of food for many animals commonly found in kelp communities.

Two principal food chains are directly associated with giant kelp — the grazers (sometimes called browsers) and the detritus feeders. Grazers, such as kelp snails, abalone, crustaceans, gastropods, and fish, feed on parts of the living kelp plant. Detritus feeders like sea cucumbers, some shrimp, bat stars, lobsters, and many crustaceans, feed on the nutrient rich decaying shed from the kelp. Some animals such as sea urchins eagerly feed on both the living plant and upon the shed, and these species are considered to be both grazers and detritus feeders. In addition, a great number of species reside in or visit kelp communities to feed directly upon various grazers and detritus feeders.

Numerous sessile (meaning non-moving or attached) animals live on the blades of the giant kelp plant. The most common of these animal groups are hydroids and bryozoans, filter feeders which colonize on the blades. Bryozoans (colonial moss animals) are often so numerous that the blades appear to be white rather than their true golden hue. Blades dominated by hydroids, tiny cousins of sea anemones, have a furry yellowish appearance. The presence of hydroids and bryozoans on the fronds attracts numerous motile populations of fishes, crustaceans, and mollusks.

Various fishes utilize the kelp for camouflage, protection, food, and as a resting place. Almost perfectly camouflaged to match the coloration of the kelp, the giant kelpfish (*Heterostichus rostratus*) and its close relative the striped kelpfish (*Gibbonsia metzi*) are fascinating to discover. Kelp fish hide in the fronds and mimic the kelp as it moves back and forth with the surge. The giant kelpfish reaches a length of 24 inches, is shaped much like a blade of kelp, and is even speckled with white dots which resemble the bryozoans that cover the blades. The smaller striped kelpfish is reddish to light brown, and is usually seen in either shallower depths where the kelp has a warmer hue or in red sea weeds along shore.

Fish such as the kelp clingfish (*Rimicola eigenmanni*) also blend in well with the kelp using the plants' protective mass as a place of hiding. The camouflage provides invaluable assistance to the clingfish by helping the clingfish capture prey while remaining undetected. The pelvic fins of clingfish are modified to a suction-like disc which enables them to hold onto the blade, a feature from which the fish derives its common name.

Halfmoon (*Medialuna californiensis*) and opaleye perch (*Girellanigricans*) feed directly upon giant kelp, while senorita fish (*Oxyjulius californica*) also ingest some kelp when they pick the surface of the blades for food. Senoritas are a type of wrasse and are often seen cleaning blacksmith fish and garibaldi in midwater cleaning stations.

Many species of midwater fishes, such as perches, also play an integral role in the kelp community. While these fish do not feed directly upon the kelp, they play important roles in various kelp forest food chains. Common residents of kelp forests, perches are planktivores, feeding upon small fishes, crustaceans, and other forms of plankton. Like other midwater fishes including jack mackerel and smelt, perches are sometimes called picker fish because these fish "pluck" or "pick" food out of midwater when feeding. In turn, perches attract larger predators to the kelp forests, fishes such as yellowtail, barracuda, and black seabass. Some of the most prominent perch species include the rainbow surfperch (*Hypsurus caryi*), striped surfperch (*Embiotoca lateralis*), and the kelp surfperch (*Brachyistiusfrenatus*).

Swimming, Snorkeling, and Diving in the Kelp

In Hollywood lore, kelp is often considered to be a man eating monster, having the ability to reach out and entangle any swimmer or diver who as much as blinks while swimming through a kelp forest. Such a reputation does make for some good, cheap late night entertainment, but it is a long, long way from the truth. With just a little common sense one can generally avoid even the slightest entanglement, but even if you do become caught in some kelp, it is quite easy to get free.

It is important to realize that a kelp plant is highly elastic. It must be in order to survive the thrashing it takes from the wave action and surge during winter storms. Without having some elasticity, the stipes would break under such a constant struggle. Being elastic allows the stipe to give with the pull of water motion and prevents healthy plants from being torn free of the bottom except in cases of severe storms.

An awareness of the kelp's elasticity is important for snorkelers, Scuba divers, and swimmers. If you do happen to become slightly entangled, or if a fin buckle gets hung up in a stipe and you find yourself towing a 30 foot long strand of kelp with you as you swim, just remember that it is easy to break the stipes by snapping them in half in much the same way you would break a pencil. But you will rarely, if ever, be able to tear the plant by stretching. Simply bending the stipe back and forth a few times should enable you to break it and avoid further entanglement. At that point you can simply remove any remaining strands without the fear of being "done in" by a mythical monster of sea lore.

If you do somehow manage to get really engulfed in the stuff, a knife can help. But be careful, you won't be the first diver who has cut through his or her own pressure gauge hose that was hidden in a clump of kelp if you manage to do so. Swimming back to a boat full of friends can be an awfully embarrassing moment after you have just sliced your own gear into pieces. I suggest trying to sneak back onboard.

Kelp as a Commercial Resource

While kelp forests are indeed a special place to fish and dive, giant kelp is useful to man in many other ways. Since 1911, when the United States Department of Agriculture sponsored a study of the kelp beds, American industries have sought economic uses of kelp. During World War I, kelp was harvested and processed for potash and acetone for use in the munitions industry. Shortly afterwards, further research discovered that algin, a colloidal substance that is a natural ingredient of kelp, has many commercial applications. Available only from certain sea plants, algin has a strong affinity for water. It is, therefore, extremely useful as a suspending, stabilizing, emulsifying, gel-producing, and film forming additive used in more than 70 commercial and household products. These include a variety of brands of ice cream, beer, fruit drinks, egg nog, candy, cake mixes, paint, paper sizing, toothpaste, and hand lotions. In some frozen foods the use of algin insures a uniform rate of thawing and provides a smooth texture. Giant kelp also contains significant amounts of potassium, iodine, and several minerals, vitamins and carbohydrates utilized as food supplements for both animals and humans.

The California Department of Fish and Game regulates the commercial harvesting of giant kelp. Modern techniques employ the use of ships which have lawn mower-like racks that are pushed through the top 3 or 4 feet of the surface canopy. After cutting, the strands of kelp are collected on large conveyor belts and taken to industrial plants for processing. Kelp related industries help provide jobs for a significant number of state residents. In fact, with all its uses as a commercial product and a recreational outlet, it was estimated in the mid 1970's that each square mile of kelp contributed over $1,000,000 to the economy of Southern California.

Protecting Our Kelp Forests

Not too many years ago man's invasion into the kelp forests placed many beds in great jeopardy. Men, both hunters and trappers, intensely overpursued the highly valued pelts of sea otters, and hunted those mammals to the point of near extinction. Sea otters prey upon sea urchins, helping to maintain a normal population size in healthy kelp forest communities. Thus, with the demise of the sea otter populations came a corresponding increase in the number of sea urchins. The unchecked expansion of sea urchins created intense competition for food among the urchins. While urchins normally prefer to feed on kelp shed — the organic debris from once living kelp — when competition for food increases they will readily forage on the kelp holdfasts which are part of the living plant. Kelp plants quickly perish when the holdfasts are eaten and the fronds are torn loose from the bottom. Once adrift, these fronds may become entangled with other kelp plants often tearing them free as well. This chain of events severely threatened the very existence of a great many kelp forest communities as thousands of square miles of kelp vanished from the sea.

In response to this problem, several governmental agencies assisted by the California Institute of Technology began implementing programs in the early 1960's to refurbish the kelp forests of Southern California. Owing at least in part to their efforts, many kelp forests made remarkable comebacks.

If anything is to be gained from this chain of events, it is the knowledge that while the kelp forest habitat appears quite rugged, it is at the same time very delicate and surprisingly vulnerable. If exploited without the proper concern, the kelp forests can be destroyed in very little time. On the other hand, if we protect this valuable ecosystem, California kelp forests can continue to provide economic resources, as well as home, food and shelter for many marine species.

Other Marine Plants

Sea Lettuce

Coralline Algae

Although there are a few noteworthy exceptions, almost all of the species of marine plants found in California belong to 3 phyla of seaweeds. They are the green seaweeds in the Phylum Chlorophyta, brown seaweeds in the Phylum Phaeophyta, and the red seaweeds in the

Phylum Rhodophyta. Plants can generally be placed into the proper phylum simply by noting their color, although determining true color can be difficult at depth.

Representatives from all three phylum contain chlorophyll, an ingredient which is 1) vital to the food making process, and 2) provides the green pigment for coloration. Brown seaweeds have an additional pigment named fucoxanthin which creates the brown hue, while red seaweeds contain a red pigment called phycoerythrin.

Other than the large brown algaes which we commonly refer to as kelp, the majority of the non-microscopic sized marine plants are rather inconspicuous. They do, however, play an important role in maintaining the stability of the substrate, in providing a food source, and in creating a place to live for numerous species.

The most prominent of the green seaweeds is sea lettuce (*Ulva lobata*), which attains a diameter of about 1 foot and bears strong resemblance to leaf lettuce.

Coralline algaes are among the most common of the red seaweeds. The cells of these plants deposit calcium carbonate which produces a hard shell-like covering, making them somewhat similar in appearance to living corals found in tropical waters. Despite this resemblance, coralline algaes are plants, not animals. They are commonly observed in the intertidal and subtidal regions throughout the state.

Two other plants that should be mentioned are eel grass (*Zostera marina*) and surf grass (*Phyllospadix scouleri*). Though they are green in color, neither is a green seaweed. Instead, both are true flowering plants, members of the Phylum Spermatophyta. Like other flowering plants, both have true root systems, leaves, stems, and produce seeds which are the result of sexual reproduction. Eel grass is the most widely distributed seagrass in North America, occurring from Alaska to Baja along the Pacific Coast and from Greenland to North Carolina in the east. Eel grass usually occurs in back bays and estuaries, while long, narrow, bright green strands of surf grass are most often seen in shallow rocky waters. The presence of eel grass tends to imply the presence of a well structured, diversified, and stabilized ecosystem. In addition to its value in preventing erosion, eel grass is a primary food source for some migratory birds as well as for a variety of invertebrates and fishes.

The Reef Community

The
Reef Community

If you want to see a host of species ranging in size from dime-sized anemones to 4-foot long horn sharks, the rocky reefs are the places to search. Unlike most tropical reefs which are built primarily from the skeletal remains of colonies of living corals, the reefs in the temperate waters of California are composed of various types of rocks. But much like coral reefs, the rock surface and the cracks and crevices in between provide ideal living quarters for a great many species. In fact, most commonly observed marine forms are associated with reef communities. California reefs are no exception, being abundant with life.

Gaining a fundamental understanding about the natural history of the various plants and animals in the reef community, and learning about their interrelationships provides fascinating insights into life in the sea. It is important to realize that the creatures found in and around the reef are quite different from the animals encountered in the open sea, along sandy bottoms, and even many species found high up in the water column directly over the reef.

Typically, the rocks are covered with several species of brightly colored anemones, sea stars, sea cucumbers, sea hares, keyhole limpets, abalone, and scallops. Black-eyed gobies (sometimes called nickel-eyed), Catalina gobies or blue banded gobies, painted greenlings, rockfish, island kelpfish, and a host of moray eels inhabit many crevices. The eels are often surrounded by cleaner shrimp which help rid the morays of parasites, dead tissue, and bacteria. California spiny lobsters are plentiful as well, at least if you look in the right cracks and caves. California lobsters lack the claws of their eastern counterparts, but divers still need quick hands to bag a limit.

The sea floor is a very important part of the reef habitat. The rocks create the perfect place for many species to rest, burrow, or to attach themselves in order to resist the sometimes violent water motion caused by surge and wave action. In addition, many life forms could not support their own weight in water if they were suspended in the water column.

These animals would perish if they were to sink to great depths in mid-ocean but the presence of shallow reefs prevents them from sinking too deep, and provides an environment where they can flourish.

Lying well within the photic zone, the rocky reefs serve as a physical barrier creating a net-like collecting surface for plankton, debris from both plants and animals, and waste matter. As a general rule, the more irregular the bottom contour, the more marine life per square yard you will discover. Many species of small marine invertebrates such as barnacles, rock scallops, anemones, and a number of species of worms are extremely well adapted for survival as long as they can find a favorable place to attach themselves. While it is true that divers often find lobster, eels, and other larger creatures under flat ledges, it is interesting to note how barren the smooth ledges are compared to convoluted reef structures where there is considerably more surface area for attachment.

Within the photic zone, debris from plant life supports the basis for a food chain of detritus feeders, animals which feed upon the debris. These animals include a number of mollusks, urchins, sea stars, and other echinoderms, worms, and crustaceans. Many of these creatures obtain the majority, if not all of their nourishment, by feeding upon the accumulation of debris.

Another parallel food chain in the reef community consists of grazers. Rather than feeding upon decayed matter, the grazers feed directly upon the various algaes such as kelp, as well as on flowering plants that are found in the reef. Kelp provides a major food source for a great many reef creatures. The existence of these 2 food chains attracts a higher level of carnivorous feeders, animals which feed upon the detritus feeders and grazers. In turn, an even higher trophic level of feeders are attracted as the cycle continues. The term trophic level refers to the stage in a food chain in which animals obtain food.

The reef communities in Northern California are less dependent upon their association with kelp than are the communities in Central and Southern California. Nevertheless, you'll discover a wide variety of animals that tend to stay close to the reef where they find protection and food. Many species of carnivorous fishes use the reef for camouflage, waiting for careless prey to swim too close. Others utilize the reef for a place to rest and hide, but are quite willing to leave the protective cover when a feeding opportunity arises.

Common Reef Inhabitants

Invertebrates

Porifera

Sponges

Cnidaria

Anemones

Corals
    purple corals
    solitary corals
Hydroids
Sea fans

Platyhelminthes and Annelida

Worms

Mollusca

Abalone
Limpets
Nudibranchs
Octopus
Scallops
Sea Hares
Navanax
Snails

Arthropoda

Barnacles
Crabs
Lobster
Shrimp

Echinodermata

Sea stars (starfish)
Brittle stars
Sea cucumber
Sea urchins

Vertebrates

Chordata

Tunicates (Sea squirts)

Cartilaginous fishes
    Horn sharks
    Swell sharks
    Leopard sharks
    Great white sharks
    Ratfish

Bony Fishes
    Moray Eel
    Rockfish
    Painted greenling
    Lingcod
    Cabezon
    Sculpin
    Kelp bass
    Black seabass
    Island kelpfish
    Ocean whitefish
    Jack mackerel
    Sargo
    White seabass
    Opaleye
    Halfmoon
    Perch
    Garibaldi
    Blacksmith
    Sheepshead
    Senorita fish
    Rock wrasse
    Wolf eel
    Blennies
    Gobies

The Reef Invertebrates

The remainder of this chapter is intended to provide a look at the natural history of many of the common reef creatures found in California waters. The animals are grouped according to their phylum, and within the phylum they are discussed in alphabetical order according to their most often used common, not scientific, name.

Phylum: Porifera

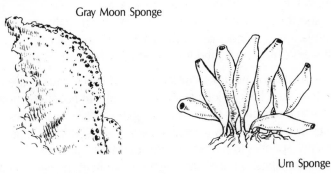

Gray Moon Sponge

Urn Sponge

Sponges

There are approximately 50 species of sponges found in California waters, but even a specialist needs a microscope to tell many of the species apart. Sponges take on a variety of shapes and colors, and while it is difficult at times to tell one sponge from the next, the presence of numerous pores all over the body makes it easy to determine that the animal in question is a type of sponge. They are almost always attached to rocks, and most are quite flexible and resilient.

In order to feed, sponges utilize tiny hair-like cilia to pump water through microscopic openings which are found all over their bodies. Sponges extract food and oxygen from the incoming water, and then expel the water, waste products, and carbon dioxide through a large opening called the osculum. In some species, such as the purple sponge (*Haliclona permollis*) the osculum are distributed in a regular pattern, while in others like the red sponge (*Plocamia karykina*), the osculum are irregularly distributed.

When speaking in laymen's terms, sponges are often described by their color or by their shape. In addition to red and purple for example, divers refer to various species of sponges as yellow (*Mycale macginitie*), sulphur

75

(*Verongia thiona*), vanilla (*Xestospongia vanilla*), gray moon sponge (*Spheciospongia confoederata*), and orange (*Tethya aurantia*). True color can be difficult to determine at depth without the use of artificial light. However, shape is another method of differentiating between species. In California waters there are urn sponges (*Rhabdodermella nuttingi*), crumb-of-bread sponges (*Halichondria panicea*), and gray and orange puff ball sponges (*Tetilla arb* and *Tethya aurantia*). The orange puff ball sponges are among the most notable at the Northern Channel Islands. But shape can also be misleading as the same species will often take on different shapes due to competition for space on a reef.

Sponges are eaten by snails, nudibranchs, and some sea stars. Some larger specimens serve as a home for shrimp, small crabs, and worms. In California water, sponges generally lack the size of tropical species, and are, therefore, not considered to be as striking by most divers.

Many species of mollusks, crustaceans, and fishes actually live the majority of their lives inside of sponges. Some, but not all, of these animals feed directly on the sponge. In some species of sponges, the spicules taste bad to many potential predators enabling the otherwise defenseless sponges to protect themselves. Made from calcium carbonate or silica, spicules are structures that help sponges maintain their shape.

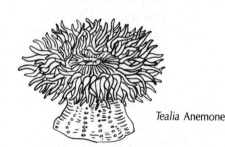

Tealia Anemone

Phylum: Cnidaria

Anemones

A simple, yet accurate, description of an anemone is "an upside down jellyfish that is attached to the bottom." Many species of anemones inhabit California reefs, attaching themselves to the rocky substrate. Some species are solitary, while others live in groups correctly referred to as clusters or aggregrations, groupings that are not correctly called colonies. Even in clusters, anemones do not share a common test or skeletal case, and are, therefore, not true colonies. Closely related to sea pansies and sea pens, anemones are described in the class Anthozoa.

They capture their prey by utilizing toxic stinging cells, called nematocysts, most of which are located at the tips of their tentacles inside of cells called cnidoblasts. The tentacles envelop the stunned prey, and with the assistance of microscopic cilia, carry the food to the centrally located mouth where it can be ingested.

Several species of California anemones are particularly striking. One of the prettiest, the Metridium anemone (*Metridium senile*) is the most commonly seen anemone in the bays of Northern California, though it is frequently seen as far south as the northern Channel Islands. While aggregations are usually dominated by a given color phase, coloration varies from pure white to brown. Metridiums are easily identified, when expanded, by the frilled appearance of the mouth. To many divers, these striking anemones are associated with the most beautiful seascapes in Northern California.

The bright red anemones (*Tealia crassicornis* and *Tealia lofotensis*) decorate many rocky reef communities, although they are more prolific the further north one explores. *Crassicornis* can be positively identified by green speckles on the stalk, while *lofotensis* is red with white spots. Though they are found in Southern California, neither species becomes prevalent until the northern Channel Islands.

The giant green anemone, or solitary green anemone, (*Anthopleura xanthogrammica*) is, as the common names suggests, both large (reaching a diameter of almost 8 inches) and solitary. The green coloration is the result of the presence of a symbiotic one-celled algae which lives deep in the tissue of the anemone. The algae requires sunlight, and solitary anemones that live in shaded areas are pale green. Mats of smaller green anemones covering the seaward walls of many pinnacles are often mistaken for *xanthogrammica*, but are really a different species (*Anthopleura elegantissima*).

Certainly no discussion of California anemones would be complete without including the aggregate corynactus anemones (*Corynactus californica*). Often these dime to quarter-sized anemones are mistakenly referred to as colonial animals, but like other anemones they do not share a common skeletal case. Large clusters often cover patches of current swept reefs so densely that the superstructure of the reefs appears to be a carpet of corynactus rather than rock. Their vivid coloration varies from bright red, to pink, to brilliant orange, to light brown, creating an array of potentially superb photographic subjects for underwater photographers.

California Hydrocoral

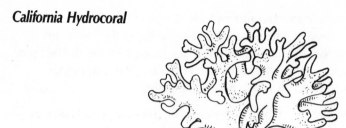

California Hydrocoral

Though California waters, like all temperate waters, lack reef building corals, we are blessed with brilliantly colored purple coral, or hydrocoral (*Allopora californica*). Clumps of this slow growing coral rarely exceed 2 feet in diameter. Like other corals, branches of purple coral are built by colonies of individual animals. In Southern California, purple coral is somewhat rare, and is most often found in deeper, current laden waters. On seamounts the clumps are often surrounded by colorful corynactus anemones and other invertebrates. In Northern California, purple coral is much more common, and is even seen at intertidal depths. Although commercially harvested for use as decor, it is illegal for sport divers to take specimens of purple coral in state waters.

Solitary Corals
(Stony Corals)

Orange Cup Coral

There are several species of solitary corals, sometimes called stony corals, in California. One that bears special mention is the orange cup coral scientifically named *Balanophyllia elegans*. This species is the most common and is often found attached to rocky substrates. While fully expanded these orange corals are about 1 inch in diameter. Though considered a solitary coral because the animals do not share a common skeletal case, it is quite common to see a number of specimens within only a few centimeters of one another. When disturbed, the tentacles retract, exposing the hard exoskeleton which is common in many anthozoans. Solitary corals feed upon plankton.

Bushy Hydroid

Hydroids

Hydroids are colonial animals which bear strong visual resemblance to small bushy plants. They usually grow in small 3 to 8-inch high feathery clumps which are attached to the surface of a rocky reef. Most appear whitish and rather inconspicuous, but do not be fooled by the look. The sting from the nematocysts can be painful. The pain does not usually last more than 30 minutes or so, but your skin may itch for several days. The stinging cells will not penetrate a wetsuit, but snorklers and divers need to be careful about brushing the colonies with their exposed skin when working in tight quarters.

There are 3 genera and many species of hydroids commonly encountered in state waters. Among them are the bushy hydroid (*Tubularia crocea*), the obelia hydroid (*Obelia sp.*), of which there are numerous species, and the plume hydroid (*Plumaria alicia*) whose stems have a feathery appearance.

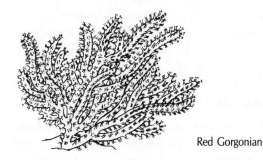

Red Gorgonian

Sea Fans

Sea fans, sometimes called gorgonians, are closely related to hydroids, anemones, jellyfish, and corals, and are therefore members of the phylum Cnidaria. Sea fans are colonies of living animals, not plants as one might suspect due to their bush-like appearance. Gorgonians form branching colonies which attach to the solid substrate. While each polyp in a sea fan is an organism unto itself, the colony shares a common test,

79

or skeletal casing. Each polyp has 8 tentacles which are expanded during feeding. Depending upon currents to provide them with the opportunity to gather food, sea fans feed by capturing tiny planktonic life forms with their extended polyps. Sea fans tend, therefore, to orient themselves with the prevailing current, and that explains why all sea fans in a given area tend to face the same direction.

The polyps are quite sensitive to light, and often retract during the day. When the polyps are extended, a diver can cause them to retract simply by touching a part of the colony. When the polyps are retracted, sea fans look like a group of twigs in the wintertime, and it is only when the polyps are extended that the gorgonians take on the bush-like look.

The polyps are equipped with stinging nematocysts which immobilize prey so that food can be drawn to the center of the polyp where the mouth is found. A number of species of nudibranchs use sea fans as hosts. The nudibranchs feed upon the polyps. In doing so, these nudibranchs are able to transfer the unfired nematocysts to their own cerata — respiratory structures on the backs of nudibranchs — where the nematocysts are used by the nudibranchs for defensive measures.

Distinguishable primarily by color, the most common gorgonians found in California are the orange (*Adelogoria phyllosciera*), purple (*Eugorgia rubens*), red (*Lophogorgia chilensis*), brown (*Muricea fructicosa*), and California golden (*Muricea californica*). All of these species are most common in the southern portion of the state though certainly red gorgonians are found as far north as Monterey Bay. Red gorgonians are quite common at the Channel Islands and at offshore pinnacles. Brown gorgonians are a mix between an off red and brown, but many specimens definitely have a reddish hue. They are, however, much less vivid red than are true red sea fans, and with a little experience it is easy to distinguish the two species.

Occasionally divers encounter a unique parasitic anemone living on the stalk of a sea fan. This species of anemone, named *Epizoanthus scotinus*, is yellow to light brown in color, bioluminescent, and most specimens are less than 1 inch in diameter. When present in dense concentrations, these parasites can kill their host sea fan.

Phyla: Platyhelminthes and Annelida

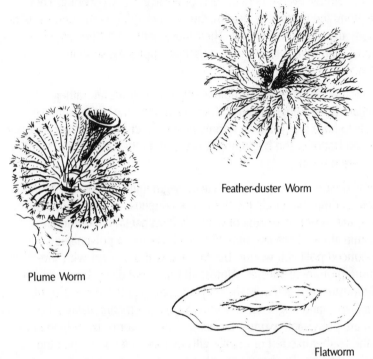

Feather-duster Worm

Plume Worm

Flatworm

Worms

In stark contrast to their terrestrial cousins, many species of marine worms are very attractive. Given that, it should not be too surprising to learn that many divers have probably admired the beauty of some marine worms without realizing that they were looking at a type of worm. Of course, there are also numerous species which are rather drab and inconspicuous. Many species of marine worms are less than one inch long but some attain a length of close to 5 inches.

There are many families of worms, and discussing all of them is well beyond the scope of this book. However, the phylum Platyhelminthes which describes the flatworms, and the phylum Annelida, the segmented worms, really should not be overlooked.

Often mistaken for nudibranchs, the bodies of many species of flatworms are brilliantly colored. Flatworms are easy to distinguish from nudibranchs if you keep two factors in mind. First, flatworms lack the obvious grouping of gills and finger-like projections found on the upper body of nudibranchs. And second, flatworms are just that — flat, lacking the body thickness of nudibranchs. Flatworms are common in mussel beds and almost everywhere there is protective rocky covering. They move along the bottom by beating the hair-like cilia on the underside of the body to create momentum. Their movement is best described as "gliding" with rhythmic waves of muscular ripples assisting in the process.

While they are relatively simple animals, flatworms are rather significant in a biological sense. They are among the most primitive animals to have true bilateral symmetry, a distinct front and rear end, and a top and bottom. The beginnings of eyes and other specialized sensory organs are present.

Annelid worms are commonly called segmented worms due to the numerous rings that circle the body. The annelids are subdivided into 3 classes, the most noteworthy of which is Polychaeta. In Southern California alone, there are more than 200 species of polychaete (pronounced pol'i ket) worms, but this text will only deal with several of the more prominent species. Almost all California divers have admired feather-duster worms (*Eudistylia polymorpha*), plume worms (*Serpula vermicularis*), and fragile tube worms (*Salmacina tribanchiata*), all of which are polychaete worms. Divers spot these worms by noticing the brilliant, feather-like red or orange gills that are extended when the animal is feeding. To observe the plumes from close range you must approach slowly and be careful not to alarm the animal, because once disturbed the worm will instantly retract its plumes. The plumes serve two functions; 1) helping the animal trap food, and 2) serving in a respiratory capacity by extracting oxygen out of the water. The rest of the animal lives in a self made tube that is burrowed into the reef.

Another species of polychaete worm, the colonial sand castle worm (*Phragmatopoma californica*) lives in large honey-comb like colonies on rocks found near sand bottoms. While the colonies live on the rocks, they depend heavily upon a constant wash of sand in order to build and maintain their tubes. These worms feed either by trapping drifting organic matter in a net of mucous, or by catching smaller particles of food in tiny hair-like structures which protrude from holes at the top of the worms' tubes. The purple hairs have a fuzzy appearance. Sand castle worms are never solitary, and divers usually spot them by noticing the large colonial construction rather than any extended plumes or feeding apparatus.

Phylum: Mollusca

Abalone

There are 8 species of abalone found in state waters, 5 of which are quite common. Abalone are members of the phylum of mollusks, a taxonomic grouping which describes a wide range of related creatures, including, scallops, octopus, squid, snails, clams, limpets, and chitons. The bodies of all mollusks are separated into 3 parts, the head, visceral mass, and foot. The head is comprised of the eyes, tentacles, and mouth; the visceral mass contains the internal organs; while the foot is large, muscular, and varies in both function and shape. In the case of the abalone the foot is used for attaching to the reef and for locomotion.

When discussing the various species of abalone, it helps to know that the circular fringe of skin bordering the foot is called the epipodeum, and the holes in the shell are properly referred to as apertures. The most important functions of the apertures are (1) to help circulate water over the gills and (2) to serve as an opening for the passage of waste matter.

Abs are further described by being part of a large class of animals commonly called gastropods. Worldwide there are over 65,000 members in this class. In California, there are over 300 different types including chestnut cowries, abalone, limpets, turban shells, whelks, topshells, and periwinkles. Members of the class Gastropoda, meaning stomach-footed, are characterized by 1) a muscular foot utilized for locomotion and for clinging to the substrate, and 2) a single shell which can be coiled, uncoiled, twisted, or in some cases internal (sea hares) or even lost (nudibranchs). The shell is secreted and maintained by the mantle which is a soft tissue found within the shell. When disturbed or alarmed, some species of gastropods withdraw into their shells and close the opening or aperture with their operculum. Abalone, however, lack an operculum and cannot display this behavior.

Like some other mollusks and a number of crustaceans, abalone are often grouped under the misleading heading of shellfish. The term shellfish is strictly a common name, and by no means connotes a scientific classification. So while abalone are commonly called shellfish, along with lobsters, crabs, and scallops, they are not true fish by any stretch of the imagination.

Abalone attach to rocky substrates with their foot, and the various species can be found in a range that varies from tide pools to about 150 feet below the surface. Although many features of the different species of

83

abalone vary only slightly, the shells vary considerably. Some species have shells that are thin, flat, and smooth, while others are thick and corrugated. Most shells have an overall cap-like shape, and appear much like the rocks upon which they live. In fact, they are often encrusted with worms. The fact that abalone blend in well with rocky bottoms makes them easy to overlook, though once you have seen 3 or 4, finding more becomes much easier.

Abalones are grazers, prefering to feed upon various forms of algae. They do so by covering the food source with their foot, and then scraping off portions with their file-like tongue, which is referred to in scientific terms as a radula. Obviously, abalone are relatively slow moving animals. They prefer to seek cover in the crevices of reefs during the day and are only seen completely away from protection at night or when food is scarce.

When properly preserved, abalone shells are used as decor. They are especially sought after for their attractive inner shell which is called the "mother-of-pearl." The inner shell has a lustrous quality and is quite colorful.

In addition to man, a number of animals like to feed upon abalone. In Northern California sea otters are their chief predators. Young sea otters can consume up to 25% of their own body weight in a day, and a full grown male can eat over 20 pounds of abalone in only 24 hours. During the latter half of the 19th century and early part of the 20th century, hunters and trappers pursued otters to the point of near extinction. Otters also feed upon sea urchins, and with the demise of the otters, sea urchin populations overran the kelp forests in many regions. With the onslaught of kelp by urchins, abalone lost their food source and abalone populations were severely threatened in many areas despite the fact that man had almost completely removed one of their most feared predators from California waters. During the same time period mankind also preyed heavily upon abalone populations as well, leading to their further demise. This sad scenario provides an excellent lesson in what happens when man radically alters the balance of nature.

A variety of other species also are likely to feed on abalone. Many fish including moray eels feed on abalone when the opportunity presents itself. In Northern California, cabezon prey voraciously on abalone populations, while in Southern California, sheephead function as a major predator. Bat rays, sea lions, crabs, and even octopi are also known to feed on abs.

Identifying the Abalone Species

Because the legal size differs from species to species, it is important that skin divers and Scuba divers be able to tell the various species apart. Though it is not as difficult as it might seem at first, if you do not do it often it is easy to forget the keys.

In common terms, the 5 species found in state waters are named the black, green, red, pink, and white. However, you should be aware that color underwater is often misleading and is unreliable as a key. Colors might not be visible, and just seeing what you believe to be pink does not mean that the abalone in question is a pink.

So how do you tell one species from the next? The following descriptions of the species should help you out:

Black abalone (*Haliotis cracherodii*): Blacks are found in significant numbers in the intertidal zone from Los Angeles to Point Conception, and as deep as 35 feet from Baja to Northern California. In Southern California, if you find an abalone in a tide pool it might be a green, but odds are you are looking at a black. The shell of a black abalone is circular, smooth, black, and has between 5 and 10 apertures. Like red abalone, the tentacles and epipodeum are dark black, but the edge of a black's shell is greenish or bluish black. Because they live in shallow areas that are constantly being pounded by waves and surf, the muscular foot develops a toughness that carries over even after blacks are prepared for serving, and for that reason blacks are generally considered the least desirable of the abalone.

Green abalone (*Haliotis fulgens*): Green abalone are commonly found in the intertidal zones of Southern California and their range extends from Baja to Santa Rosa Island. Preferring shallow water, they are commonly found in less than 20 feet, and rarely at a depth of more than 60 feet. Greens can easily be identified by virtue of their gray or light green tentacles. The edge of the shell is usually rather thin and reddish brown in color. Large greens reach diameters of more than 9 inches, with the olive green to brown shell being oval in shape. There are normally between 5 and 7 slightly raised apertures that are often connected by a dark red band.

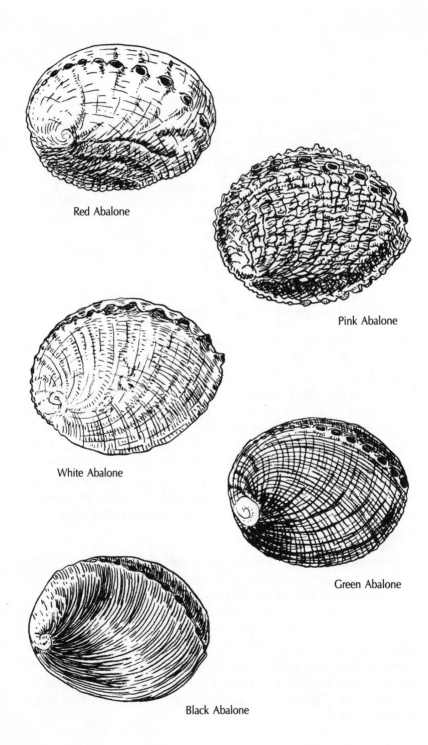

Red Abalone

Pink Abalone

White Abalone

Green Abalone

Black Abalone

Red abalone (*Haliotis rufescens*): Attaining diameters of greater than 11 inches, reds are the world's largest abalone. The range of reds extends from Baja to Oregon. South of Point Conception, reds are normally encountered between 20 and 100 feet, while they are commonly found in the intertidal zone to the north. Rarely are reds taken in water that is warmer than 60° F; thus, the further south you dive, the deeper you normally have to go to find red abs. These gastropods are characterized by their thick, dark black tentacles and the bright red margin along the growing edge of the shell, though some reds have green, pink, or even white margins. Red abs usually have 3 to 5 slightly elevated apertures, though as with most species, some older specimens possess fewer apertures.

Pink abalone (*Haliotis corrugata*): Pink abs inhabit reef areas between San Miguel Island and Baja from the intertidal zone to depths up to 200 feet. They are the easiest of the abalone to identify due to the pronounced, corrugated design along the edge of the shell and their dark black tentacles. Reaching widths of more than 8 inches, the top of the shell is also corrugated, but this feature is often hidden by attached growth or erosion. Pinks have between 2 and 4 open holes in the shell.

White abalone (*Haliotis sorenseni*): Whites, sometimes called Sorenson abalone, are the favorite of many ab hunters, being noted for their delicious and tender meat. Having a range that extends from San Miguel Island to Baja, whites prefer deeper, colder waters. In Southern California white abs are often found only at depths in excess of 100 feet, while in northern waters they are found as shallow as 35 feet. Whites are characterized by yellow tentacles and a bright orange foot. The outer edge of the shell is usually red or pink, and is normally rather thin. The shell is highly curved or arched, thick, and circular, with 3 to 5 obviously elevated apertures. White abalone attain diameters of up to 9 inches.

Other Species

There are 3 other species of abalone that are occasionally seen in California waters. They are the threaded (*Haliotis kamtschatkana assimilis*), the flat (*Haliotis walallensis*), and the pinto (*Haliotis kamtschatkana kamtschatkana*). Together these species comprise less than 1% of the total population of California abs.

Sex in the Sea

Many lower forms of invertebrates reproduce simply by breaking up their bodies in a process called asexual reproduction. In asexual reproduction, an entirely new organism originates from a part of an already existing specimen, and there is no need for different organisms to provide sperm and eggs for fertilization. Sponges, many cnidarians, and some echinoderms can reproduce in this manner. If reproduction can occur without sex, a logical question for the scientific community to pursue is: "why does sex occur at all?" What is the advantage, if any, of sexual reproduction?

Surprising to many, the advantage is not all that well understood. Genetic recombination, a term which describes sexual reproduction through cross-fertilization, is both a destructive and constructive evolutionary process. Scientists are well aware that the more highly specialized an adaptation is, the more unlikely that genetic recombination will improve upon the end result, and the more likely the recombination is to prove to be disruptive. It is, therefore, no accident that asexually reproducing plants are more common in extremely harsh habitats, such as the polar regions. The survival ability of these plants has been established, and the process of genetic recombination is more likely to lead to problems for the species, rather than to enhance its ability to survive. So now you know the disadvantage to reproduction by genetic recombination. As for the advantage ... well, according to scientific experts the advantage probably is deep rooted in a mechanism through which damaged chromosomes have a chance to be repaired.

Whatever the combination of events that led to sexual reproduction, in an evolutionary sense the consequences have been monumental. No genetic process creates diversity as rapidly or as profoundly.

Sexual reproduction first evolved among marine organisms. The eggs of most fishes are produced in batches by the female of the species. When the eggs are ripe they are released into the water. The eggs are not harmed by seawater, and are capable of being fertilized by combining with sperm that has been released in the water by the male.

The effective union of egg and sperm in the sea creates a number of problems for marine organisms that utilize external fertilization. Predation upon the rather defenseless stages of immature forms of many specimens is intense, so the number of young must be large if the species is to survive. Both eggs and sperm become rapidly diluted after being released into the water, so timing is a critical factor. In sea urchins, oysters, and a variety of additional invertebrates the chance for successful spawning is greatly increased by the presence of chemical substances known as pheromones. These invertebrates reproduce by releasing sex cells into the water, but the release is not a random event. Present in both sperm and eggs, pheromones are detected by other nearby members of the same species, and that detection induces the others to release their own sperm and eggs into the water.

Fishes have adapted 3 different strategies of reproduction. In most species fertilization and development of the eggs is external. In these cases the eggs contain only enough yolk to sustain the developing fish for a short period of time, after which the juvenile must fend for itself. Thousands of eggs are often fertilized, but only a handful of these fish manage to survive long enought in the demanding aquatic environment to reach sexual maturity. The development of externally fertilized eggs is comparatively rapid.

In some species fertilization is internal, the male injecting sperm into the female. Biologists describe three different methods of internal fertilization. They are oviparous, ovoviviparous, and viviparous. Oviparous development occurs when the fertilized eggs are laid outside the mother's body where they complete development. In ovoviviparous reproduction the eggs are retained inside the mother where they hatch and the young are then immediately released into the water. In viviparous reproduction, the eggs of the undeveloped young hatch within the mother where the young obtain further nourishment as development proceeds. The young are later released, being fully developed at birth. Blue sharks and hammerhead sharks are examples of species that reproduce through the strategy known as viviparous reproduction.

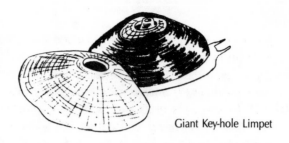

Giant Key-hole Limpet

Limpets

Limpets are snails, members of the class of gastropods, and in a biological sense are closely related to abalone. The shells of limpets are quite simple in construction, lacking the spiral design found in many gastropods. Herbivorous animals, limpets are usually seen on rocks where they graze upon algae which they scrape off the rocks with their rasp-like teeth or radula.

There are several species of limpets found in California waters. Some of the most common are giant key-hole limpets (*Megathura crenulata*), volcano limpets (*Fissurella volcano*), file limpets (*Acmae limatula*), owl limpets, (*Lottia gigantea*), fingered limpets (*Acmaea digitalis*), and kelp limpets (*Acmaea insessa*).

Because they are so often misidentified as abalone by beginning divers, probably the most notorious of the limpets are key-hole limpets. These animals attain a shell diameter of 5 inches and an overall body length of up to 8 inches. The mantle can be black, white, grey, or a mottled combination and extends over the shell. The design created by the opening at the apex of the shell is the root of the name, key-hole. Key-hole limpets are not as heavy or as hard as abalone. Although abalone and limpets do not possess similar markings because of their overall shape and preference for rocky habitats, it is easy to understand how beginning divers mistake limpets for abalone. Though eaten in Japan, limpets are not considered a food source in California, and many new divers have experienced an embarrassing moment when asking an "old pro" what to do with their limpets.

Tiny kelp limpets normally get no larger than ¼" in length, but they are often seen on the flat center portion of a stipe of feather-boa kelp. Kelp limpets eat their way into the kelp making a small pit, not quite as large as the animal itself. On almost any feather-boa plant, divers can either find the limpets or a series of pits.

90

Nudibranchs

Nudibranchs are shell-less mollusks which are sometimes called sea slugs. Meaning "naked gills" in Latin, the term nudibranch refers to their exposed respiratory organs. Trying to describe a nudibranch to anyone who has never seen one underwater is a nearly impossible task, especially without the use of an accompanying photograph or drawing. Words alone simply do not do them justice as nudibranchs are considered by many to be the most beautiful of marine animals. Ranging in length from less than ½" to close to a foot, nudibranchs occur in a wide diversity of striking shapes and colors. In fact, in California waters alone there are more than 160 species.

The family Nudibranchia is subdivided into 4 suborders, the dorids, aeolids, dendronotids, and arminids. The distinguishing characteristics of each suborder is beyond the scope of this text, but by paying close attention to the general body shape, the finger-like projections on the body called cerata, and external sensory organs called rhinophores, it is possible for knowledgeable divers to quickly place nudibranchs in their respective suborder.

Some nudibranchs blend in well with their surroundings, while the coloration of others makes them "stand out in the crowd." Nudibranchs are slow crawlers and generally poor swimmers, making one wonder how they survive in the sea. The answer lies in the fact that they simply do not taste good to very many potential predators. In fact, their only known major predators are navanax (*Aglaja inermis*), who readily feast on their close cousins, but then again, navanax also prey upon each other.

Nudibranchs are grazers, using their radula (a rasping tongue-like organ) to secure food. The radula are species specific, being designed to help each type of nudibranch secure its favorite foods.

All nudibranchs are hermaphroditic — having both male and female sex organs — though self-fertilization is rare. Most species display their hermaphroditic quality throughout their lives, while others change from males to females with age. Nudibranchs lay their eggs in a characteristic mass, whose shape and form is unique to each species.

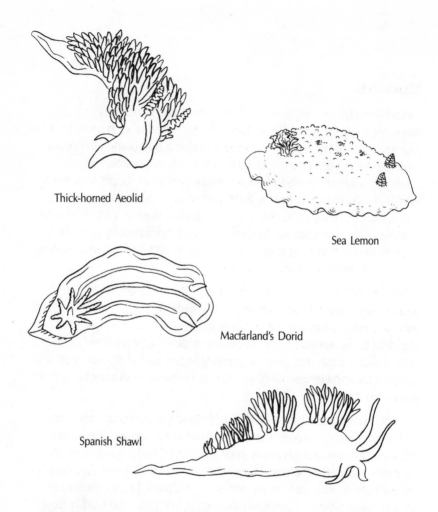

Thick-horned Aeolid

Sea Lemon

Macfarland's Dorid

Spanish Shawl

While there are a great many beautiful nudibranchs, several common species are especially striking. They are the Spanish shawl (*Flabellinopsis iodinea*), the thick-horned aeolid (*Hermissenda crassicornis*), pugnacious aeolid (*Phidiana pugnax*), Macfarland's dorid (*Chromodoris macfarlandi*), Hopkin's rose (*Hopkinsia rosacea*), California dorid (*Hypselodoris californiesis*), white-lined dirona (*Dirona albolineata*), three-color polycera (*Polycera tricolor*), the Santa Barbara nudibranch (*Antiopella barbarensis*), and the rainbow dendronotid (*Dendronotus iris*). One of the largest and most commonly seen species is the sea lemon (*Ansidoris nobilis*), whose color varies from burnt orange to deep yellow. Mentioning only these species is certainly not intended to slight any others, as all are truly magnificent animals.

Sea Hares and Navanax

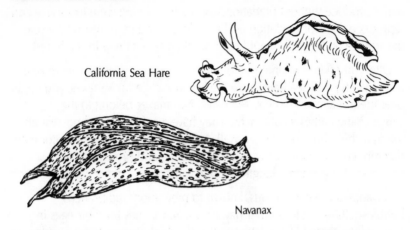

California Sea Hare

Navanax

Sea hares inhabit many California reefs. The 3 most commonly observed species differ significantly in either color or size. The largest species, *Aplysia vaccaria*, can be distinguished by its solid black color and can reach a length of 2 feet or more. *Vaccaria* ranges from Morro Bay to the Coronado Islands. Another species, *Apylsia californica*, is characterized by its splotched purple and brown body. Full grown specimens of *californica* are usually only slightly smaller than *vaccaria*. The range of *californica* extends from Humbolt Bay, Alaska to the Sea of Cortez. A third species, *Aplysia taylori*, also has a splotched purple and brown body, but it can be distinguished by its smaller size. When full grown, these sea hares are less than 8 inches long. *Taylori* inhabits reefs between British Columbia and San Diego.

Sea hares are gastropods, being closely related to nudibranchs. They lack external shells, but do possess a thin internal remnant. Like octopi, sea hares eject a purplish ink when threatened or roughly handled. It is interesting to note that like their close cousins, the nudibranchs, sea hares are hermaphroditic, meaning each individual has both male and female reproductive organs.

They are, however, incapable of self fertilization and often mate in large groups. Sea hare eggs appear as an entangled mass of long yellow strings and are common sights on reefs. The mass of eggs is often as large as a basketball. The number of eggs is very high, and studies have shown that some sea hares produce as many as 95 million eggs a month for months at a time. The young hatch in 10 to 12 days, and are subject to heavy predation by plankton feeders. In fact, scientists have calculated that if just one adult layed its normal load of eggs and all were to survive and reproduce without predation, and their offspring were to survive and reproduce without predation as the cycle continued, in just over a year the surface of the earth would be more than 6 feet deep in sea hares!

Often mistaken for large nudibranchs, a navanax (*Navanax inermis*) is actually a type of sea hare. In fact, when not called navanaxes, you might hear them referred to as bay sea hares. Navanaxes belong to the subgrouping called tectibranchs. They have a soft body and possess an internal shell, but no external shell. Navanaxes are variable in color with different combinations of white, yellow, blue, and brown being most common. They reach a length of up to 7 inches.

Navanaxes are carnivorous, known to prey upon nudibranchs, bubbleshells, as well as other navanaxes. Navanaxes lay their eggs in long, white, thread-like gelatinous strings which appear to be wadded up like a tangled ball of yarn. The egg masses can be found in reef areas throughout the year.

Chestnut Cowrie

Snails

There is an abundance of species of snails in California waters. Many species are drably colored and rather inconspicuous, and normally the animal living within the shell is almost completely hidden from view. However, if you cautiously approach, you will usually be able to see the foot, eye stalks, and tube-like mouth called the proboscis. Some species also display a colorful mantle, a flap of skin that is exposed when covering the shell. Secretions from the mantle are used to maintain the shell, giving a healthy shell its quality of luster.

Of all the snails seen in California waters, perhaps the most striking is the chestnut cowrie, *Zonaria spadicea*. While their range does extend throughout California waters, they are much more numerous in Southern California than in Central and Northern California. Attaining a length of just over 2 inches, the surface of the shell is a combination of white, tan, and chestnut brown.

The shell of a healthy specimen has a highly polished, glossy appearance caused by seceretions from glands in the mantle. The mantle is a soft tissue that is used to maintain the shell of all gastropods. However, in many species the mantle is either unexposed or not as striking. The golden mantle of the chestnut cowrie is highlighted by many black spots. If the mantle is exposed upon first sighting the cowrie, and you wish to take a closer look, be careful not to disturb the animal. When alarmed the mantle is quickly retracted.

Two other species of cowries are found in Califronia waters. They are the little coffee-bean (*Trivia californiana*) and the large coffee-bean (*Trivia solandri*). Both species are considerably smaller and less colorful than chestnut cowries, are also found in California waters.

Several other commonly seen species that inhabit California reefs are the Kellet's whelk (*Kelletia kelletii*), the blue top shell (*Callistoma ligatum*), the wavy top turban (*Astraea undosa*), the red turban (*Astraea gibberosa*), the California cone shell (*Conus californicus*), the 3 wing murex (*Pteropura trialata*), several species of olive shells in the genus *Olivella*, and several species of bubble shells in the genus *Haminoea*.

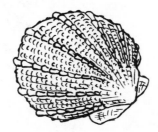

Rock Scallop

Scallops

Many rocky reefs from Baja to British Columbia are inhabited by rock scallops (*Hinnites multirugosus*). These mollusks are bivalves (in the class Pelecypods), meaning they have two shells called valves which are hinged together and are joined by a tough, yet flexible ligament. The soft body lives inside of the shell which can be tightly closed when the animal feels threatened.

In their juvenile form, rock scallops are free swimming, drifting in currents in a planktonic stage. After they reach about 1 inch in size, the scallops attach themelves to rocks by secreting a limestone-like substance which secures the lower shell to the rocky reef where the animal will live its entire adult life.

As adults, rock scallops feed upon drifting plankton, and, as might logically be expected, are usually found in areas where heavy current and strong upwellings are common. They are almost always horizontally oriented and strongly prefer areas that are shaded from direct sunlight. Rock scallops can be found from the intertidal zone to depths of approximately 180 feet, and commonly attain diameters of up to 5 inches, though 10 inches is not unheard of.

When adult rock scallops feed, they expose their colorful mantles and eyes. The eyes are well developed and serve as a defense mechanism by sensing changes in light intensity. These scallops are best spotted by sighting the colorful mantle, as the shell is almost always encrusted with both plants and animals. Scallops with bright orange mantles are males, those with the brown to green mantles are females. Rock scallops are preyed upon by octopi, starfish, and many species of drilling snails. Humans find rock scallops to be a tasty favorite as well.

Two-spotted Octopus

Octopus

While there are at least 8 species of Octopi reported in state waters, positive identification in the field proves extremely difficult. In fact, even members of the scientific community have problems determining who's who among the various species. This is due to the fact that different species have astonishingly similar characteristics in their adult forms. It is interesting to note that as adults the various species are quite similar despite the fact that some octopi have a planktonic stage that lasts several weeks, while other species are immediately benthic upon hatching.

Two species bear special mention in a discussion of California reef areas. They are the smaller, and much more common, two-spotted octopus (*Octopus bimaculatus*) and its larger cousin, the giant octopus (*Octopus dolfleini*). (The commonly observed sand dwelling species

California Reef Creatures

, A diver poses for a photograph with ever curious aribaldi.

2. Male rock scallop surrounded by corynactus anemones.

. California spiny lobsters typically inhabit crevices in rocky reefs.

5. Wolf eel.

4. Moray eels surrounded by cleaner shrimp.

6. Blood star on purple coral.

7. Moray eels normally shy away from divers, but they can often be lured out of the reef with an offer of food.

Both the red *Tealia* anemones (above left) and the Metridium anemones, also called white-plumed anemones (above right), are common in the cooler waters of Central and Northern California.

Great white sharks are rare in California. Here a diver enjoys the photographic opportunity of a lifetime!

10. Lingcod.

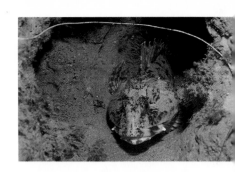

11. Cabezon.

12. Female sheephead.

13. Island kelpfish.

14. Male rock wrasse.

15. Catalina or blue-banded goby.

16. Ratfish.

17. Ronquil.

. Opaleye.

19. Treefish.

0. Gopher rockfish.

21. Black-and-yellow rockfish.

2. Rosy rockfish.

23. Olive rockfish.

24. Vermillion rockfish.

25. Copper rockfish.

26. Spanish shawl nudibranch.

27. Pugnacious aeolid nudibranch.

28. Sea lemon.

29. Egg mass of sea lemon.

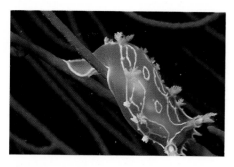

30. Tritons nudibranch (also Festive triton).

31. Navanax.

32. California sea hare.

33. Sea hare egg mass.

. Encrusting sponge.

35. Orange cup corals.

. Red sea fan.

37. Close-up of feeding sea fan polyps.

. Green abalone.

39. Giant keyhole limpet.

. Chestnut cowrie with mantle extended.

41. California cone shell.

42. Snakeskin brittlestar.

43. Soft sea star (also fragile rainbow star).

44. Knobby sea star (also giant-spined sea star.)

45. Orange sun star.

46. Giant red sea urchin.

47. California sea cucumber.

48. Feeding barnacle.

49. Lacy bryozoan.

Octopus rubescens is covered on page 146, in its appropriate habitat section.) The two-spotted octopi rarely get larger than 2 feet across, and are distinguishable by the prominent dark spot under each eye. These octopi probe the reef at night, seeking a diet of clams, crabs, and an occasional small fish. When a two-spotted octopus captures a clam, it will either try to forcefully open the shell to obtain meat, or it can choose to bore a hole into the clam's shell with its strong beak, before injecting a paralyzing venom that allows the octopus to pry the bivalve apart.

Octopi rely upon the "jet engine" design of their mantle for swimming. By opening the mantle and then contracting it forcefully, octopi gain thrust. They generally prefer to move by walking along the bottom on their arms, which are lined on the underside by two rows of sucker cups.

Moray eels are a primary predator of octopi. Octopi emit clouds of "ink" when threatened by morays or other predators, not so much as a visual "smoke screen" as is often thought, but rather to attempt to dull the olfactory senses of the eels as the octopi swim away. Octopi utilize several additional methods of defense as well. They are capable of changing their coloration, the texture of their skin, and their shape within seconds as a camouflage technique.

Nocturnal animals, octopi are frequently seen by night divers. But it is not uncommon to locate an octopus resting at the mouth of its den during daylight hours, especially if the diver is thorough in an examination of the cracks and crevices in rocky areas. The den is often surrounded by a mound of empty shells which are the discarded remains from prey. This pile of debris near the den is called a midden.

The giant octopus, which is so often found in water off of the west coast of Washington and British Columbia, is frequently seen in Northern California, but is only rarely encountered as far south as the Channel Islands. While giant octopi attain weights up to 150 pounds in the waters of the Pacific Northwest, the largest specimens inhabiting California rarely get larger than 30 or 40 pounds. Big or small, these octopi are very docile animals, and are not at all like the monsters of Hollywood sea lore. In fact, almost all octopi tend to be extremely shy, and will typically retreat upon sighting a diver.

Phylum: Arthropoda

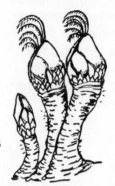

Goose-neck Barnacles

Barnacles

For many years barnacles were scientifically associated with mollusks, but in 1830, a British biologist pointed out the many similarities shared with crustaceans. Like lobsters, crabs, and other crustaceans, barnacles develop from an egg into a larval stage. In their larval stage, barnacles are free swimmers, traveling high up in the water column until they find places to settle and become bottom dwelling organisms.

Once settled, barnacles are sessile animals, attaching themselves quite firmly to the substrate, whether it is a rocky reef, a pier piling, the bottom of a ship, or the skin of a whale. Most species are hermaphroditic.

The two most notable groups of barnacles in California waters are the goose-neck barnacles (*Pollicipes polymerus*) and several species of acorn barnacles that make up the family Balanidae. Goose-neck barnacles are often associated with mussel populations and are the species of barnacle that so often cover pier pilings. A goose-neck barnacle possesses a long, fleshy "neck," or stalk, with which it attaches to the bottom. It was once believed by a prominent Greek naturalist in the mid-to-late 1500's that geese actually hatched from these barnacles. Modern science has come a long way indeed!

reach out into the water column. The feet are repeatedly drawn through the water in a rhythmic, sweeping motion in an attempt to catch tiny food particles. The feet are then drawn into the shell where the food can be ingested.

To the boating crowd, barnacles are often considered no more than a major nuisance. But to divers, the delicate intricacies of the feeding behavior of barnacles can be truly fascinating to observe and photograph.

Acorn barnacles, especially the giant acorn barnacle (*Balanus nubilus*) can be distinguished from close relatives by their fleshy lips which lack the bright colors found in some other species. In addition, acorn barnacles are not found within the intertidal zone. Giant acorn barnacles are often found on exposed, current swept seamounts. As adults, all barnacles depend upon water movement to bring them food, as the animals are "cemented" to the substrate. Barnacles acquire food by capturing it with their cirripedia or "feather feet." The feet are extended from the shell through the mantle opening, and spread apart as they

California Spiny Lobster

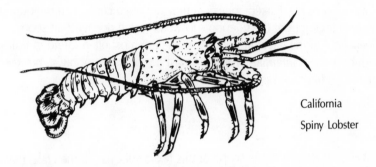

California
Spiny Lobster

Lobsters are crustaceans in the class Crustacea and the order Decapoda, as are crabs and shrimp. But whether the California spiny lobster (*Panulirus interruptus*) is in fact a lobster is a point of some debate. Unlike Maine lobster, the California spiny lobster has no large, pinching claw, though the females possess a small pincer on the last pair of walking legs. Due to the lack of claws, many people refer to the spiny lobster as a crayfish, although it is biologically speaking, quite different from freshwater crayfish.

The largest California spiny lobster ever documented weighed just over 35 pounds and measured 3 feet long. Known to inhabit the cracks and crevices of rocky reefs from Point Arguello to Baja, it is believed that these crustaceans live up to 100 years or longer. As juveniles, lobsters shed their shells, or molt, several times a year. Even so, they grow only a little more than 1/4" per year. As they grow older and larger, the frequency of molting declines.

To reach sexual maturity, a lobster must survive a minimum of 7 years. Spawning occurs from May through July. Before the female lays her eggs, the male deposits strings of sperm packets on her abdomen. When her eggs are laid, the female digs a hole in the packets, and the eggs are fertilized. The eggs are carried in grape-like clusters on the small abdominal legs called swimmerets, which are possessed only by females. The eggs hatch in about 10 weeks.

Upon hatching, the larvae are planktonic, drifting at the mercy of ocean currents for up to 8 months before settling to the bottom where they live the rest of their lives as benthic creatures. As adults, lobsters feed on a wide range of sources including snails, sand dollars, other shellfish, and decaying plants and animals.

Exactly where the best place to find lobsters of legal size is a hotly debated issue among California divers. At different times over the course of most years, legal sized "bugs" as they are often called can be found in depths ranging from the tidepools to water much deeper than sport divers go. During daylight hours, lobsters are most often found in holes, with only their antennae projecting past the opening. At night, however, they leave the recesses of the reef to scavenge.

Crabs

While it is beyond the scope of this text to discuss all the crabs that inhabit the rocky reef communities in California waters, several species bear special mention. Certainly that is the case with the hairy hermit crab, *Pagurus hirsutiusculus*. Like all hermit crabs, this species has a well armored forward body, but a very soft abdominal section. Such a build could create some serious problems, but these clever crustaceans protect their weakness from potential predators by residing in a shell that has been discarded by a snail, usually a type of turban. Only the well armed claws and the front of the body remain exposed. It is quite difficult to remove a hermit crab from a shell, as the abdomen is curved to accommodate the spiral construction of most snail shells. Hairy hermit crabs reach a length of close to 4 inches.

Four other groups of rock dwelling crabs should be noted here. They are the porcelain crab (*Petrolisthes cinctipes*), the northern kelp crab (*Pugettia producta*) which is sometimes called the kelp crab, the southern kelp crab (*Taliepus huttallii*), and red crabs (*Cancer productus*). Porcelain crabs are characterized by their flat, smooth bodies and the porcelain-like appearance of the body carapace which is typically about 2 ½ inches in diameter. Unlike most true crabs, this species has only one pair of antennae between the eyes, instead of two.

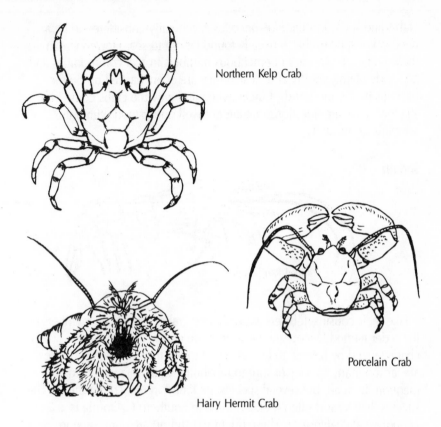

Northern Kelp Crab

Porcelain Crab

Hairy Hermit Crab

The coloration of northern kelp crabs varies from olive green to brown. These crustaceans attain a width of 6 inches when the legs are included, and are commonly seen hiding in the stipes and holdfasts of giant kelp. Northern kelp crabs can be distinguished from southern kelp crabs by the fact that in northern kelp crabs the diameter of the claws is much larger than the diameter of the legs, while the diameter of the claws of the southern kelp crab is not as large as that of the legs. In addition, the body of a southern kelp crab is longer than it is wide, and the rostrum is obviously pointed.

Red crabs are distinguished by their bright red color and claws with black tips. Like all cancer crabs, the front margin of the carapace is made up of many serrated tooth-like projections. Red crabs reach a diameter of 6 inches and are considered a highly valuable food source. When using the common name, it is easy to confuse this species with the species that is scientifically named *Pleuroncodes planipes*, which is referred to by a number of common names including red crabs, pelagic red crabs, tuna crabs, and squat lobster. *Plueroncodes* is generally considered to be a deep water species that is usually found far out to sea. However, at times these crabs are present in tremendous numbers in reef communities and on sandy plains, shortly after which they are often washed up on the beaches by the thousands. Exactly when and why the crabs come into shallow water are questions that are not well understood within the scientific community.

Shrimp

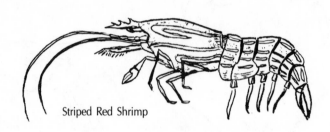

Striped Red Shrimp

Like their cousins, lobsters, crayfish, and crabs, shrimp are members of the order named Decopoda. Decopods are crustaceans that are characterized by having 10 legs attached to the thoracic region. There are far too many species of shrimp inhabiting California waters to mention them all, but several species are found in reef communities. The species that is most often seen by divers in Southern California is *Hippolysmata californica*. These bright red shrimp are often seen in aggregates, gathered around moray eels. The eels and the shrimp share a symbiotic relationship classified as mutualism. The shrimp benefit in two ways; 1) by feeding upon parasites, bacteria, and dead tissue that they find on the eels, and 2) by being protected from potential predators by the mere presence of the moray, as the morays do not feed on the shrimp. The eels benefit from the shrimp by virtue of the fact that the shrimp clean the eels.

From Monterey north, coon stripe shrimp (*Pandalus danae*) are commonly seen in intertidal areas. The colorful red body is decorated with a series of bright blue stripes. Coon stripe shrimp reach a length of close to 3 inches.

Abalone shrimp, *Betaeus harfordi*, are commonly found living under the shell of abalones. These shrimp are usually a glossy brown to purple, but solid black as well as solid blue specimens are not uncommon. The shrimp benefit by having a home, but whether the abalone are harmed, benefitted, or are simply unaffected is still being debated.

Phylum: Echinodermata

Sea Stars or Starfish

In California, numerous species of starfish inhabit both rocky reefs and sand bottoms. All starfish are in the class Echinodermata, as are brittle stars, sand dollars, sea cucumbers, and sea urchins. In fact, all echinoderms — approximately 6,000 species — are marine animals.

Most echinoderms are characterized by having short, calcareous spines which usually project out from the top surface of the body, and all species possess a water vascular system, a feature that is unique to this phylum. As is immediately obvious in the case of most starfish, the bodies of mature specimens are typically symmetrical in fives (pentamerous radial symmetry) or multiples of fives. In the case of starfish, most have five arms radiating out from a central disc. Many echinoderms possess a series of tiny organs called pedicellariae, which upon close examination look like miniature pliers and serve a protective function, helping to keep the body surface free of debris.

The class Asteriodea includes sea stars, while brittle stars are grouped together in the class Ophiuroidea. Perhaps the best known of the California starfish is the knobby star (*Pisaster giganteus*) sometimes called the giant star. The white spines of giganteus are surrounded by randomly positioned bright blue aureole. Reaching a diameter of 18 inches, the knobby starfish is similar in appearance to the ochre star (*Pisaster ochraceus*). Ochre stars are the more common of the two species. The colors of the spines of ochre stars are more variable, ranging from yellow, to orange, brown, blue, purple, and red. Ochre sea stars can be easily distinguished from knobby stars by the fact that the colorful aureole of ochre stars are arranged in distinct networks.

Like many species of sea stars, those just mentioned feed heavily upon mussels. There are two somewhat contradictory theories that describe the way by which sea stars force mussels to open their shells. One theory states that when eating a mussel, the starfish grasps opposite sides of the shell with its tube feet overpowering the mussel gradually by pulling on each half of the shell, eventually forcing the shell to partially open. The second theory claims that the sea star forces the mussel closed in order to prevent the mussel from obtaining oxygenated water. When the oxygen supply becomes too low, the mussel forces its shell open, and the starfish is able to get to the meat.

When eating, the starfish either turns its own stomach inside out and places it against the shell of the mussel near the ligament that joins the shell in order to weaken the shell; or immediately begins to ingest the meat by forcing its stomach through an opening of less than 1/100 of an inch and enveloping the meat with digestive tissues. Many bivalve mollusks have irregularities in their shells that create natural openings of 1/100 of an inch. Once inside the mollusk the sea star secretes digestive enzymes and begins to digest the soft tissues of its prey.

Sea stars prey heavily upon many bivalves, some crustaceans, and decaying organic matter. Their mouth is located in the center of the lower surface of the body. Sea stars are heavily preyed upon by sea otters.

There are several species of sea stars that possess more than five arms. The six-rayed star (*Leptasteria hexactis*) has, as the name suggests, 6 arms. A group of sea stars commonly called sun stars have between 8 and 24 arms. The species of sun stars most often encountered are the orange sun star (*Solaster stimpsoni*), the sunflower star (*Pycnopodia helianthoides*), the morning sun star (*Solaster dawsoni*), and the rose star (*Crossaster papposus*).

Other starfish which are commonly seen in a reef community are bat stars (*Patiria miniata*), soft stars (*Astrometis sertulifera*), red stars (*Mediaster aequalis*), fragile stars (*Linckia columbiae*), and vividly colored blood stars (*Henricia leviuscula*). Bat stars are characterized by the webbing between their short, thick arms. Their coloration is variable though they are typically red on the upper surface, with a yellowish underside. The upper surface ranges from purple, to green, to red, to a mottled combination. A bat star is extremely well adapted for its life on the reef. They will eat almost anything. When feeding along the bottom, a bat star extrudes its stomach outside of its body, and then absorbs and digests organic debris.

Soft stars are characterized by their long, flexible, soft arms and long orange and blue spines. Compared to other starfish, soft stars are rather fast travelers, able to traverse up to 10 feet in an hour.

Red stars are colored a bright red, and have a series of small plates that border the arms. Red stars are easily distinguishable from red colored bat stars, as red stars lack the characteristic webbing between their arms — a prominent feature of bat stars. The arms of red stars are also considerably thinner. Red stars are commonly seen on encrusting coralline algae in northern waters.

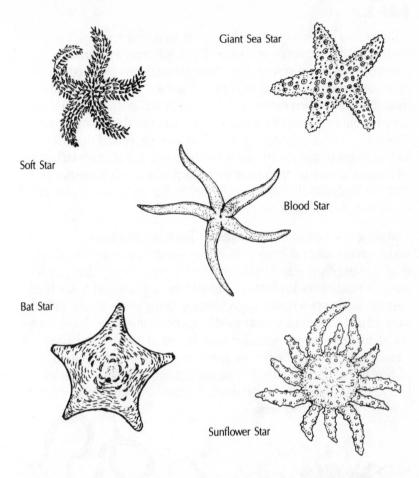

Giant Sea Star

Soft Star

Blood Star

Bat Star

Sunflower Star

Fragile stars are a mottled red and gray, and are unusual in that their thin arms are often of different lengths. These sea stars stand out for two reasons. First, while all starfish are able to regenerate lost parts, the fragile stars complete the process the fastest. Second, a lost arm may develop into a complete animal, though perfect symmetry is rarely obtained. In most species of sea stars at least 1/5 of the central disc must be attached to the arm before regeneration is possible, and even then the process can take up to a year. As is the case with fragile stars, perfect symmetry is seldom achieved when regeneration is required.

Blood stars are a vivid orange to red. The slender, smooth arms lack both spines and pedicellaria. Interestingly, in this species, the female carries the eggs until the young are hatched, rather than depositing the eggs on the reef and crawling away, as is the case with most species.

Brittle Stars

Brittle stars are so named for their rather delicate arms, which tend to break off at the slightest disturbance. If lost, the arms are quickly regenerated from the central disc. These echinoderms are to be distinguished from sea stars. Brittle stars can be identified by (1) their long, thin arms which radiate from a small central disc, and which are waved about in a snakelike fashion, (2) the fact that their arms are not "webbed" — instead being sharply set off from the central disc, (3) the lack of tube feet that starfish use for locomotion, and (4) their lack of an ambulacral groove on the underside of each arm which is present in sea stars. (An ambulacral groove is a channel on the underside of each arm in sea stars that contains the tube feet.)

Many species are found in California, but telling them apart underwater is rather difficult as individuals of the same species often display markedly different coloration. Colors typically vary from a wide range of solid colors to mottled combinations. It is unusual to see brittle stars out in the open during daylight hours. Being very sensitive to both sunlight and the artificial lights used by divers, brittle stars almost always spend their days in well secluded areas buried in the sand or hiding under rocks. Some brittlestars are detritivores, preferring to feed upon particulate matter which collects on the bottom, while other species feed primarily upon suspended microplankton which they capture with their tube feet and long arms.

Red Sea Urchin

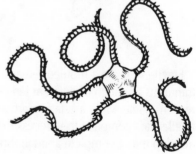

Banded Brittle Star

Sea Urchins

All sea urchins are members of the phylum Echinodermata, and like sand dollars, are in the class Echinoidea. In addition to the 5-sided symmetry displayed by other echinoderms, echinoids are further characterized by having movable spines and tube feet.

106

There are 5 species of sea urchins that are commonly seen in California waters, 3 of which are seen in rocky reef communities. They are the giant red urchin (*Strongylocentrotus franciscanus*) which varies in color from bright red to black, the purple urchin (*Strongylocentrotus purpatus*), and the green urchin (Strongylocentrotus drobachensis) which is found in Northern California. The other two species are sand dwellers. (See page 153.)

Giant red urchins are the largest, with the test reaching a diameter of up to 7 inches, while the most abundant are the purple urchins. Purple urchins rarely attain a diameter of more than 3 inches. They frequently occur in dense populations in kelp forests and among larger boulders on exposed coasts in the intertidal zone. Green urchins are very easy to recognize due to their short spines.

As swimmers and divers we tend to think of urchins as animals with whom contact is to be avoided whenever possible. Urchins are well known, and intensely disliked, due to the pain caused by their long, sharp spines. Fortunately for those of us who frequent California waters, the spines of urchins found in state waters do not usually cause the severity of pain caused by those found on many tropical urchins, though a puncture wound can be quite painful. The spines are mucous coated and wounds often become infected.

Urchins often dig their way into a sand bottom underneath a rocky ledge. Over time they can grow so large that the urchins can not escape from the original opening and must resort to suspension feeding instead of grazing.

Sea urchins play a significant role in the overall ecology of the kelp forest and reef communities. They are preyed upon by sea otters and several species of fish which bump them off the rocks, and then bite into the vulnerable, exposed tube feet. Urchins are grazers, feeding on a variety of algaes. In healthy kelp forests, urchins feed primarily on the shed of kelp, but they are also known to feed upon the holdfasts of living plants. Urchins use a complex 5-toothed structure in their mouths known as Aristotle's lantern when eating. The teeth come together in a manner similar to the way a bird's beak closes. The teeth, which can be extended or withdrawn into the mouth as a unit, are quite strong, allowing the urchin to scrape algae off of rocks and to capture small fragments of organic matter.

Sea Cucumbers

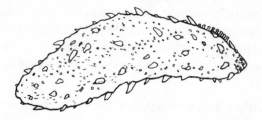

Common Sea Cucumber

I can not count the number of times I have heard a relatively new diver surface and ask, "what are those things that live on the bottom and look like cucumbers? I must have seen at least 25." Well, by common name, the answer to the just posed question is "sea cucumbers." Scientifically speaking, sea cucumbers are classified as echinoderms, being in the class Holothuroidea. The two most commonly seen species of sea cucumbers in California waters are the Southern California cucumber, often called the common sea cucumber (*Stichopus parvimensis*) and the sweet potato cucumber (*Molpadia arenicola*). Sea cucumbers can generally be described as elongate or sausage shaped. They are soft-bodied, flexible, and the skin of some species (*parviminsis*) is covered by warts and soft spines.

Both of these species reveal several interesting adaptations. When relaxed their bodies are soft and pliable, but when threatened they become shorter, thicker, and much harder by contracting their muscles and ejecting water from various body tissues. Most sea cucumbers can eject their sticky internal organs when threatened, and though they have few natural enemies it is believed that the ejection can be used to help ward off predators. Some species of sea cucumbers demonstrate a remarkable ability to live in areas that are severely oxygen deficient.

A smaller species of California sea cucumber is the white sea cucumber (*Eupentacta quinquesemita*), an echinoderm whose design looks somewhat like a centipede. Whites, which are colored white to light orange, reach a maximum length of about 4 inches and have long non-retractable spines covering their bodies.

Sea cucumbers play an important role in many reef communities by feeding upon detritus. In California, sea cucumbers feed primarily on kelp shed.

108

The Reef Vertebrates

Phylum: Chordata

Tunicates (Sea Squirts)

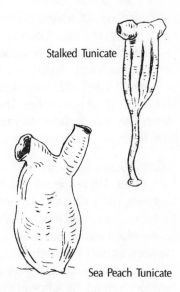

Stalked Tunicate

Light Bulb Tunicates

Sea Peach Tunicate

No matter how many times one reads in a biology text that tunicates are primitive vertebrates, after seeing them in the water, one has to refer back to the text just to be sure. Yes tunicates are, in fact vertebrates, comprising the subphylum Urochordata. The confusion lies in the fact that as divers we see the adult forms in reef communities. It is only in their larval stage that tunicates have a notochord and a central nervous system, characteristics that make scientists classify tunicates as vertebrates. Tunicates feed by pumping water through their bodies in order to remove small particulate matter, and from that behavior have received their commonly used group name, sea squirts.

Tunicates can be either solitary like the stalked tunicate (*Styela montereyensis*), or colonial like light bulb tunicates (*Clavelina huntsmani*). There are several species commonly seen in California, and in their adult form, they look a lot more like a terrestial plant than they do any type of vertebrate. As the common names (light bulb tunicate, sea peach, stalked, and elephant ear to name a few) imply, tunicates come in a variety of shapes. Most species are white, yellow, or orange.

The scientific names of some commonly seen species are as follows: sea pork (*Amaroucium californica*), sea peach (*Halocynthia aurantia*), glassy (*Ascidia paratropa*), and elephant ear (*Polyclinum planum*).

Fishes

There are at least 551 species of fish found in the Pacific Ocean in California's coastal waters. However, in taxonomic terms there are only 3 major classes of fishes; Chondrichthyes, Osteichthyes, and Cyclostomata. The class Chondrichtyes describes the approximately 850 species of cartilaginous fishes, sharks, rays, and skates. The class Osteichthyes includes all fishes with bony skeletons. Worldwide, there are more than 17,000 species of bony fishes. The vast majority of fishes in California, and in waters throughout the world, possess skeletons made of bones. Bony fishes are sometimes referred to as teleost fishes.

Cartilaginous fishes are believed to have evolved much earlier than bony fishes. Those species which lack both bony skeletons and jaws, the hagfishes, slime eels, and lampreys, belong to the class named Cyclostomata (round mouth). While many of these species are thought to have evolved earlier than many cartilaginous species, in California all members of the class Cyclostomata are either very rare or they inhabit extremely deep water. Extensive discussions about these species are therefore beyond the scope of this book, as these fish are rarely encountered by recreational enthusiasts.

While physiological differences other than the composition of their skeletons are evident when comparing cartilaginous and bony fishes, both laymen and scientists categorize fishes according to those terms. The following chart points out some basic differences between bony fishes and cartilaginous fishes.

	Bony Fishes	Cartilaginous Fishes (Sharks, Rays, and Skates)
Scales	present in most species; usually large, rounded, and bonelike in composition	no true scales, but do possess dermal denticles which are fine tooth-like structures
Gills	usually one on each side of the head covered by a bone-like operculum, which is used to pump oxygenated water over the gills	usually possess 5 to 7 gill slits on each side of head; slits lack a covering or operculum
Air bladder (a gas filled sac that enables fish to maintain neutral buoyancy at any depth in water)	typically present	absent
Method of Reproduction	generally by spawning, young most often hatch from eggs	by copulation in all species, young of most species born live, but there are numerous exceptions
Anatomy	mouth usually in terminal position at front of head; tail usually symmetrical, with backbone ending where tail starts; teeth in jaw sockets; skulls are segmented	mouth usually below head; tail typically asymmetrical with vertebrae extending into upper lobe of tail; teeth not firmly affixed to jaw; non-segmented skull

Author's Note: Most fishes described in this book are associated with the reef community. Because there is such an abundance, I have grouped the species according to their class, either Chondrichthyes—the cartilaginous fishes, or Osteichthyes—the bony fishes. Within their class the species are grouped into their respective family. The family name is listed first by the popularly given family name, and second by the scientific name which appears in parentheses.

CARTILAGINOUS FISHES: Sharks and Ratfish

BULLHEAD SHARKS

(Family: Heterodontidae)

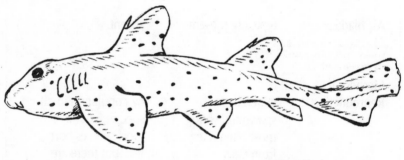

Horn Shark

Horn Shark

Of the more than 350 species of sharks found worldwide, only a few are commonly seen in the reef habitat of California waters. However, one species that is often seen resting on the rocks from Baja to Point Conception is the horn shark (*Heterodontus francisci*). So named for the extension of a white spine-like appendage (actually a modified dermal denticle) which protrudes from the dorsal fin, horn sharks are characterized by their short, blunt heads, pursed lips, and distinct ridges above their eyes. The bodies are tan with black spots. These sharks appear to be more closely related to catfish than to sharks, but in scientific terms horn sharks are in fact true sharks.

Horn sharks reach lengths of up to 4 feet, though it is quite common to see specimens that are less than 2 feet long. They are docile, bottom dwelling creatures that feed upon crustaceans, mollusks, and small reef fish. Primarily nocturnal feeders, horn sharks are equipped with two distinct types of teeth. The front teeth have sharp cusps used for grasping prey, while the rear teeth have rounded cusps used for crushing.

When encountered during daylight hours, horn sharks are usually seen resting on the bottom. Though they are generally considered to be quite harmless, horn sharks can make dramatic photographic subjects.

Divers often find horn shark egg casings resting on the rocky bottom or slightly entangled in kelp. The screw shaped eggs are about 5 inches long, and have a leathery look. Colored dark brown, they possess prominent spiral flanges which make the eggs readily identifiable.

CAT SHARKS
(Family: Scyliorhinidae)

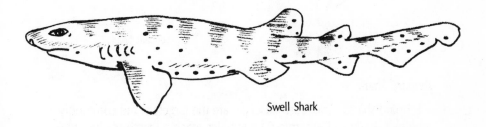

Swell Shark

Swell Shark

Next to horn sharks, the most commonly seen shark in California reef communities is the swell shark (*Cephaloscyllium ventriosum*). Swell sharks are found from Monterey southward. Swell sharks are so named due to their ability to swallow large amounts of water and/or air when threatened. Doing so allows the sharks to increase their size and helps them wedge themselves into cracks, making it difficult, if not impossible for their natural predators to dislodge them. Like horn sharks, swell

113

sharks are docile, bottom dwelling, nocturnal animals. Preferring to spend the day nestled away in caves or crevices, swell sharks pursue a diet of small reef fish at night.

Their maximum length is about three feet. Swell sharks can be distinguished by their yellowish brown bodies which are covered with brown spots, their flattened heads, and wide mouths. Although swell sharks are generally thought to be solitary creatures, they do congregate in some areas at given times.

SMOOTHHOUND SHARKS
(Family: Triakidae)

Leopard Shark

Leopard Shark

Leopard sharks, *Triakis semifasciata,* are the largest of the commonly seen reef sharks in California. The females reach a length of close to 6 feet, while the males rarely exceed 3 feet. Both sexes have rather slender brownish-gray bodies speckled with prominent black crossbars and black spots. The diet of leopard sharks consists of mollusks, crustaceans, and some reef fish.

Like other California reef sharks, these animals are extremely docile, and usually flee upon sighting a diver. When breeding, concentrations of 50 or more of these sharks are occasionally seen in shallow water just beyond the surf zone. During those times the sharks show little, if any, reaction to the presence of swimmers or divers. These sharks are also believed to breed in back bays and estuaries.

114

MACKEREL SHARKS
(Family: Lamnidae)

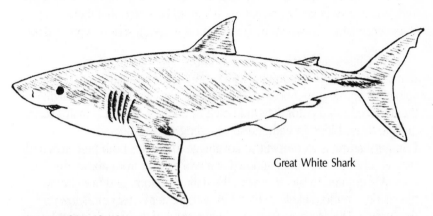

Great White Shark

Great White Shark

Great white shark . . . white shark . . . referred to as the white pointer or white death in Australia and New Zealand. In scientific circles the animal is known as *Carcharodon carcharias.* No matter what you call it, the name alone conjures up an image of terror in the minds of many experts and laymen alike.

And yes, it is true, great white sharks inhabit California waters. They have done so for years. Considering that fact and the fact that there are so few attacks on divers, surfers, and swimmers, it seems that far too much has been written about their predatory capabilities. However, considering their size, or just from catching a glimpse of one underwater, it is equally logical that one could never really write enough.

Despite all of the highly publicized research that has been conducted on white sharks in the past decade, realistically we don't know a lot of what there is to know. There are plenty of theories, a great deal of educated speculation, some ongoing research projects, but a lot of unknowns remain. It is likely that our fear of white sharks is more closely related to media exploitation and humanity's natural fear of the unknown than it is to the true nature of an animal whose natural history is not well understood.

In terms of their basic biology, a significant amount of information about great white sharks does exist. It is in the area of behavior where knowledge is most lacking. Great white sharks can be generally described as having a heavy, spindle-shaped body, with a moderately long, conical snout. Their bodies can roughly be described as looking like enormous overpressurized footballs. White sharks possess large (as long as 3 inches), flat, serrated, blade-like teeth, comparatively long gill slits, and a very large first dorsal fin. It should be noted that there is no direct correlation between tooth size and the overall size of white sharks.

Great whites are black to slate gray above, and have whitish underbellies. Some specimens display one or more small white splotches on the upper body and tail. Being dark on the top and light on the underside — a color pattern known as countershading — helps many pelagic fishes blend in with the surrounding water colors. Countershading is a competitive advantage that helps both predators and prey go undetected. Looking down at a white shark from above, the animal is colored to blend in with the darker bottom, and from below the sharks are more likely to blend in with lighter water colors toward the surface. This is not to say that a white shark will go undetected by potential prey, but countershading does provide predators with the edge of perhaps going unnoticed a little longer than they might have otherwise.

The tail of white sharks is described as being almost homocercal or lunate, meaning the tail is nearly symmetrical, with the upper and lower lobes being the same shape and size. Most species of sharks possess obviously asymmetrical tails with the upper lobe being noticeably larger than the lower lobe. With most white sharks the upper lobe is slightly, but only slightly, larger than the lower lobe. Homocercal tails are characteristic of many of the fastest swimming fishes, including mako sharks and all members of the tuna family. The body of a great white shark is equipped with a pair of prominent, strong caudal keels which provide both added strength to their tail fin and increased stability. The caudal keels are located on either side of the body about 18" in front of the tail fin where the body is its narrowest.

While scientists are quick to point out the construction of the tail as a key to identification, in a wilderness setting it can be rather difficult to focus your attention on that end of the animal. If you are able to do so, you are likely to think the tail is disproportionately large. That is true when compared to the tails of many other fishes, and is a good indication of the fact that white sharks are capable of rapid bursts of speed.

Although great white sharks have been found in waters all over the world, they are believed to prefer to inhabit the waters near coastal areas and offshore islands of continental shelves in temperate seas. Certainly some large specimens have been taken in gill nets miles from the nearest land, but it is generally accepted that while some white sharks venture forth into deep ocean basins, breeding populations only exist close to shore. Great whites are well known to penetrate shallow bays in coastal waters of continental shelves. They are also known to congregate near offshore islands that have large populations of seals and sea lions. While sightings are generally considered rare, the known range of great white sharks includes a large percentage of tropical and temperate waters around the world.

The natural size attained by great white sharks is certainly a major factor contributing to Hollywood's ability to create and perpetuate the menacing reputation that surrounds white sharks. Taken in waters off the coast of Cuba, the largest documented great white reached a length of 21 feet and weighed 7,100 pounds. While those numbers are certainly impressive, it is their girth that makes white sharks appear so overwhelming. It is no exaggeration to state that with the record sized white shark, or even one half that size, a very large person with exceptionally long arms could not reach around the shark and join hands either one foot behind the snout or one foot in front of the tail. The record for a great white shark taken in California water is 18 feet 2 inches long, weighing just under 5,000 pounds.

I once saw a 16 foot long great white that had been caught by fishermen being lifted out of the water by a winch. The first attempt to raise the shark broke the winch. When lifted by a second winch, four harbor seals fell out of the shark's stomach. Full grown harbor seals weigh up to 300 pounds. That scene made a lasting impression on me, and should help you gain some insight into just how large an animal a great white can get to be.

At birth white sharks are believed to be between 3 and 4 feet long, but scientists are not positive about this point. Four foot long specimens have been taken in some shallow back bays and estuaries, where pupping is believed to occur, but it is really not known how large or small white sharks are at birth. Small specimens are caught by fishermen in certain shallow back waters and bays on a predictable basis leading scientists to believe that well defined pupping grounds do exist. Located just north of San Francisco, Tomales Bay is considered as a primary pupping grounds in north central California.

The Art of Locomotion in the Ocean

Superb swimming ability in various marine creatures is a function of the successful combination of shape, locomotive movements, and various adaptations.

Two factors are always working against speed in water. They are "drag" and "pressure resistance." Drag is a result of four factors — (1) the shape of an animal, (2) the speed with which it moves, (3) the surface of the creature, and (4) the nature of the flow of water over that surface. Tests prove that the faster the speed of travel, the greater the drag. Pressure resistance is a function of shape and is the consequence of having to part and push water in front of a moving body and force it to flow around the body. The smoother the flow along the body of a moving animal, the less resistance, and the faster the animal can swim. In general, the more streamlined and longer the body, the smoother — or more laminar — the flow. Fishes that are torpedo shaped, such as barracuda and wahoo, can, therefore, attain the fastest speeds.

Many bony fishes such as garibaldi and rockfishes are greatly assisted in their ability to overcome the effects of gravity by the use of a gas-filled swim bladder. A swim bladder allows fish to increase their size, thus increasing the amount of water displaced, without increasing their weight. This adaptation allows the fish to easily achieve a state of neutral buoyancy. Fishes that lack a swim bladder — such as sharks and rays — sink if they do not swim constantly because their combined skeletal and muscular weight is heavier than the weight of displaced sea water.

Fishes such as angel sharks and bat rays spend a considerable portion of their lives hiding or resting on the bottom. These animals utilize their weight to help anchor themselves on the bottom. In most cases, they are able to remain motionless even when encountering current and surge. For many open ocean species such as blue sharks, the lack of a swim bladder forces them to swim for their entire lives without the opportunity to stop or rest.

Even a brief consideration of the various shapes of marine animals will tell you quite a lot about where animals live, how they feed, and how they protect themselves from predators. A look at the variety of shapes found in fishes will help illustrate that point. The basic fish shapes are:

118

1) *Fusiform:* More or less torpedo shaped, with a slightly rounded head and a long, thin, tapering body as seen in blue sharks, mako sharks, and barracuda.
2) *Laterally compressed:* Flattened from side to side as in the case of garibaldi.
3) *Dorsoventrally compressed:* Flattened from top to bottom, thus creating a wide, flat profile. Angel sharks, halibut, and bat rays have bodies that are dorsoventrally compressed.
4) *Attenuated:* Long and thin, having a body diameter which is very small when compared to the length. Moray eels are classic examples of fishes that have an attenuated shape.
5) A small selection of unusual shapes, each with only a few representatives such as seahorses and ocean sunfish.

Most species possess the combined characteristics of more than one category. Each species has evolved over the years to adapt characteristics which yield the best chance for survival in the particular manner in which they live.

Many of the fastest swimming species — tuna, open ocean sharks, and barracuda — are generally fusiform, but the fusiform shape has been altered by a slight flattening from side to side. Their flattened fusiform shapes and reinforced, powerful tails combine to create an excellent design for the speed needed by open water predators.

Fish with laterally compressed forms are extremely manueverable, but are less capable of generating the speeds attained by fishes with more fusiform shapes. In reef areas that are filled with a myriad of hiding places, laterally compressed species such as garibaldi are well designed for slipping into thin crevices to seek prey or to hide from predators when threatened.

Dorsoventrally compressed representatives such as bat rays and angel sharks are capable of maintaining a low profile and moving almost unnoticed along the bottom where they typically blend in so well. Functioning without a swim bladder, bottom dwelling rays have evolved a unique method of locomotion through the evolution of large, powerful pectoral fins. And when resting on the bottom, the additional surface area created by the enormous pectorals enables rays to get a better "grip."

The general consensus among experts is that it takes white sharks approximately 10 years to reach sexual maturity. A ten year old shark will be in the neighborhood of 14 feet long and can weigh between 2,000 and 3,000 pounds. Studies indicate that immature specimens strongly prefer temperate waters, while mature white sharks are more likely to travel to tropical waters and to venture into deep water.

White sharks are presumed to be ovoviviparous, meaning pregnant females produce shelled eggs which develop inside the body of the mother. These eggs contain a considerable yolk sac but the young sharks do not receive nourishment directly from the female in the form of a placenta. Like many, but not all species of sharks, whites are suspected to practice uterine cannibalism. This means that while several eggs hatch, the strongest of the young sharks devours its brothers and sisters before birth. It is believed, though not documented, that white sharks produce only one or perhaps two young when they pup. The gestation period is thought to be rather long, 12 to 13 months, but again these numbers are the result of educated speculation as no scientist has ever had the opportunity to observe a pregnant female over any extended period of time.

As is the case with many species of sharks, whites appear to be sexually segregated during most of the year. Exactly when, where, and how they mate is a matter of some speculation. Females often display mating scars, the result of biting from males during copulation. In many species of sharks, the male is believed to bite the female on or near the pectoral fin in order to get a grip on the female before twisting underneath in an effort to insert a clasper in the act of mating. The skin of female white sharks, as is the case with many sharks, is up to 3 times thicker in the pectoral region than over the rest of the body. This adaptation is believed to prevent the females from getting injured while mating.

The combination of their massive size, powerful jaws, large razor sharp teeth, and their ability to accelerate rapidly help make white sharks true superpredators. They are known to prey upon a wide variety of bony fishes such as sturgeon, salmon, halibut, hake, rockfish, lingcod, cabezon, mackerel, and tuna. White sharks also readily feed upon small reef sharks, some larger sharks including basking sharks, sting rays, eagle rays, dolphins, elephant seals, harbor seals, sea lions, and even some birds such as gulls and gannets when the opportunity presents itself. It is generally accepted that mature white sharks prey rather heavily upon marine mammals, while smaller specimens enjoy better success stalking rays and bottom fish.

It is interesting to note that the remains of sea otters have never been found in the stomach contents of white sharks, although they are well documented to inhabit the same waters. And while several species of seals are preyed upon rather heavily by white sharks, California sea lions are not. In fact, through 1985 there were only 2 documented attacks on California sea lions by white sharks. The reason that so few California sea lions have been taken by white sharks is thought to be that these sea lions are so highly maneuverable, and simply put, are better swimmers than white sharks.

In recent years the public has witnessed some excellent cinematography showing white sharks biting a number of metallic objects including swim steps and shark cages. The biting scenes gave many people the false impression that white sharks will eat anything. A more accurate analysis of facts indicates that the biting is probably related to the electrical field given off by the metallic swim steps and shark cages when they are submerged in salt water. Scientists know that sharks can detect electrical fields up to 10,000 times better than other animals. All animals constantly emit an electrical field. For millions of years the repsonse of biting anything that produced an electrical field has produced food for sharks. In essence, the introduction of an unnatural object such as a metal shark cage into the sharks' environment has confused or tricked the shark into biting an object that emits an electrical current.

White sharks maintain a core body temperature that is usually between 3 and 8° F above the surrounding water temperature. Considering their elevated body temperature, their lack of an insulating layer of fat, and the fact that their extremely large liver is so active, many scientists feel rather certain that white sharks must eat fairly often, or at least in large quantities.

Not much is known concerning the population dynamics of white sharks. Few have ever been tagged and recovered at a later date, so information concerning their migration or preference for territoriality is merely speculative. White sharks generally appear singly, although they are occasionally observed in pairs. And when a large amount of food is readily available, whites are known to congregate. Smaller white sharks are generally observed to give way to larger animals when they are competing for food.

RATFISH

(Family: Chimaeridae)

Ratfish

In California, ratfish (*Hydrolagus colliei*) are observed much more often the further north you explore. Once purported to be a missing link between cartilaginous fishes and bony fishes, scientists no longer believe that to be true. Ratfish are one of the few members of the Chimaeridae family, and do have primarily cartilaginous skeletons.

Attaining a maximum length of just over 3 feet, ratfish can be distinguished by their lack of scales on the body, weak skeletons which are quite pliable, and a strong venomous spine at the origin of the dorsal fin. Unlike most cartilaginous fishes ratfish do possess a gill cover called an operculum, and they do not breathe through a spiracle as is the case with many bottom dwelling rays, skates, and sharks. The upper side of the body is usually a metallic bronze with white spots, while the underbelly is silver. In overall shape, ratfish look much more like common reef fish than they do sharks or rays. A pair of claspers, part of the males' reproductive system, are however, evident upon close inspection. Claspers are present in the males of all cartilaginous fishes and are a characteristic which distinguishes cartilaginous fishes from bony fishes.

In many reference sources, ratfish are reported to be common. While that might be true at depths greater than sport diving limits, they are only rarely seen in less than 100 feet of water in Southern California. However, they are seen rather frequently in relatively shallow water in the northern part of the state.

BONY FISHES

EELS

(Family: Muraenidae)

Moray eel

California moray eels (*Gymnothorax mordax*) are common residents of many shallow reef communities south of Point Conception. Surprising to many, the dark green eels are scientifically classified as fish, although several significant differences between true fishes and eels are readily apparent even to the novice biologist. Moray eels lack the large gill covers possessed by true fishes. The gill covers are used by true fishes to pump water over the gills so they can extract oxygen from the water. The lack of gill covers requires morays to use their mouths as a billows type pump to continuously circulate a fresh supply of oxygenated water across their gills. Hence, morays are constantly opening and closing their mouths in an effort to breathe. Doing so has led to a lot of bad publicity in the media and misrepresentation by Hollywood filmmakers. Almost always photographed with mouths agape, exposing their fang-like teeth, morays are often portrayed as monsters of sea lore. Nothing could be farther from the truth. In fact, upon first sensing a diver, morays often withdraw into their holes.

Reaching lengths of close to 6 feet, morays are primarily nocturnal feeders, preferring to feed upon small reef fish and octopi. With the olfactory regions of their brains being well developed, these fish rely heavily upon their sense of smell to detect prey. The lack of gill covers, and their lack of large, paired fins enables morays to back up without getting hung up, as they skillfully manuever about the reef. Another significant adaptation is their mucous coated skin which is considerably different than the scaly skin of most reef fish. This layer of lubrication also adds to their increased maneuverability in tight quarters and to their skill as reef predators.

Morays are often surrounded by, and at times covered by, cleaner shrimp. In a classic symbiotic relationship called mutualism, both the eels and the shrimp benefit from their relationship. The eels do not eat the shrimp as the shrimp clean parasites, bacteria, and dead tissue from the eels, even entering the mouth at times. The shrimp have a ready food source, in addition to being protected from potential predators by the eel's presence. (See page 102 for additional information about the relationship between moray eels and cleaner shrimp.) Bluebanded gobies (*Lythrypnus dalli*), sometimes called Catalina gobies, and less noticeable zebra gobies (*Lythrypnus zebra*) are also known to rid morays of parasites.

Wolf-Eel: Wolf-eels are not true eels. They are more closely related to blennies. For information about wolf-eels, see page 135.

ROCKFISHES
(Family: Scorpaenidae)

Rockfish

The family Scorpaenidae contains more species than any other family of fishes found in the entire Eastern Pacific, including 62 species which are endemic to California water. By common names, all rockfish (genus, *Sebastes*) and some, but not all, fish that are commonly called sculpin are members of the Scorpaenidae family. As a general guideline, rockfish are bass-like looking fishes, characterized by the pronounced projection of the lower jaw, large eyes, prominent spines in front of the dorsal fin, and the bony support extending down from the eyes and gill slits. Individual specimens can be difficult to identify in the field because they go through so many color phases, especially as adults. In addition, the color patterns, shape, and number of spines changes markedly with age in some species. Most rockfish rate highly with spearfishermen, sport fishermen, and commercial fishermen. When fully grown, most specimens are less than 18" long.

The common name for the family, as you might suspect, comes from the fact that as adults, rockfish frequent rocky habitats. Many species that are seen only near the deeper reefs of Southern California are commonly seen in less than 30 feet of water in the northern part of the state.

All Sebastes bear live young. For the first few weeks of life, the hatchlings swim near the surface in the open sea, often miles from the closest reef community. Juvenile rockfishes are heavily preyed upon in mid ocean, and often seek refuge in patches of free drifting kelp.

Rockfish feed upon a wide variety of mollusks, crustaceans, and smaller fishes. In turn, they are preyed upon by rays, seals, sea lions, and bottom dwelling sharks.

Not only do rockfish taste great, but they make excellent photographic subjects as well. Framed against a colorful reef, the dramatic faces and rich color patterns combine to produce strong photographic possibilities. Though all have potential, several of the most photogenic species as adults include vermilion rockfish (*Sebastes miniatus*), black-and-yellow rockfish (*Sebastes chrysomelas*), gopher rockfish (*Sebastes carnatus*), rosy rockfish (*Sebastes rosaceus*), starry rockfish (*Sebastes constellatus*), treefish (*Sebastes serriceps*), and china rockfish (*Sebastes nebulosus*). The juveniles of many rockfish are much more striking than the adults.

Why Some Fish School

Most of the thinking concerning why fish school is based on educated guesses. Presently, four theories as to why fish school are generally well accepted. First, there is the concept of safety in numbers. If the school encounters a predator, perhaps the predator will eat fish "b" and "c," but fish "a" will survive. A predator can eat only to its feeding capacity, and if the schools are large enough it is unlikely that all specimens will be taken. If that is the case, the odds that some specimens will reach sexual maturity and be able to reproduce are increased, insuring the propagation of the species. The preservation of the species is of utmost importance.

A second theory centers around the idea of confusing a predator that is a more capable swimmer. When the predator encounters the school, the overwhelming number of fish present makes it difficult for the predator to decide which fish to pursue. During the moment of hesitation, perhaps all the potential prey can escape, or the predator will become so confused that it will simply take aim on the general direction of the school and not single out a given fish. Failing to single out prey almost always proves to be a costly mistake for predators.

Third, it is a well accepted concept that, in general terms, big fish eat little fish. Some scientists believe that a large school might appear as a single large fish or large animal to a potential predator, and the "single" animal will appear larger than the predator. The potential predator, therefore, considers itself to be potential prey and leaves the school alone.

Fourth, schooling also serves to keep reproductively active members of a population in close proximity to one another. Since most fishes reproduce by external fertilization, the chances for successful union of sperm and eggs are greatly increased if mature females and males are able to receive signals that help them to simultaneously release eggs and sperm into the water.

GREENLINGS
(Family: Hexagrammidae)

Painted Greenling

Often called convict fish by sport divers due to the 6 or 7 alternating dark and light vertical stripes along a body which resemble convicts' apparel, the painted greenling (*Oxylebius pictus*) inhabits rocky reefs throughout the state. Despite their numbers, painted greenlings are rarely taken by sport fishermen. That phenomenon is likely due to their small mouths. Generally described as cigar-shaped, convict fish reach a maximum length of 10 inches, but most are 6 inches or shorter. These fish will often display a curious nature, if divers remain relatively still.

Lingcod

Though they are members of a different family (*Hexagrammidae*), lingcod are similar in appearance to cabezon (family *Cottidae*). Both cabezon and lingcod have elongated bodies and rather large heads. Cabezon, however, lack scales, while lingcod lack the frilled facial markings found in cabezon. With the exception of salmon, lingcod are the most popular ocean sport fish from Northern California to Washington. Bottom dwellers as adults, lingcod vary greatly in color from grey-brown to green or blue. Most are spotted or mottled on their upper bodies. Lingcod prefer rocky ledges and are found from intertidal depths to 1400 feet. As a rule, the farther north you go, the bigger the lingcod. The largest lingcod ever documented in California weighed 41.5 pounds, while in British Columbia the record is 105 pounds. Though naturally tinted green prior to preparing, the meat is excellent.

Sea lions are the natural predators of lingcod, while the lings' diet consists almost exclusively of fish. Tagging studies have verified that lingcod are quite territorial.

SCULPINS
(Family: Cottidae)

Cabezon

Another of the "must mention" species is the cabezon (*Scorpaen ichthys marmoratus*), which is found from Alaska well into Baja. Cabezon are notorious for their unusual shape. They have bulbous heads, stumpy tails, goggle eyes, tufted "eyebrows", and very large, fanlike pectoral fins. The females are mottled green, and the males are mottled with patches of brown and red.

The male cabezon guards the nest until the eggs hatch. At that time the hatchlings drift with currents for several weeks until by random coincidence they settle in shallow water and begin their benthic life. Cabezon are known to reach a length of 3 feet and weigh up to 25 pounds, but most are considerably smaller.

Cabezon catch their prey by stealth. They are referred to as a "lay and wait" predator. Cabezon blend in well with rocky surroundings, and they simply rest on the bottom and wait for unsuspecting prey to come too close. When the prey, usually a small fish, gets within range, cabezon dart out and capture it.

Sculpin

There are more than 45 fish in California waters that are popularly called sculpin of one type or another. These fish are grouped together in the family Cottidae. One fish that is not included in this family is the fish most commonly called "the sculpin (*Scorpaena guttata*)" by laymen. The use of the term "the sculpin" illustrates a classic case of the importance of communicating by use of scientific names within the scientific community. *Scorpaena guttata* is a member of the family Scorpaenidae, which also includes all rockfish.

Commonly seen species from the Cottidae family include the tidepool sculpin (*Oligocottus maculosus*), the rosy sculpin (*Oligocottus rubellio*), the wolly sculpin (*Clinocottus analis*), and the red Irish lord (*Hemilepidotus hemilepidotus*). All of these species experience a variety of coloration, with splotches of red, purple, yellow, orange, white, green, and brown. All are less than 7 inches long, and distinguishing one from the next is often difficult at first.

CLINIDS
(Family: Clinidae)

Island Kelpfish

Island kelpfish (*Heterostichus rostratus*) are curious residents of the many rocky reefs in Southern California. Reaching five inches in length, island kelpfish have a grayish-red body with approximately six dark bars vertically arranged down the side of their body. The dorsal fin is usually a brilliant reddish-orange. When first approached by divers these fish appear timid and will quickly retreat into a crevice in the reef. However, if you are patient, odds are that within a minute or so the fish's sense of curiosity will rule and it will dart out of the hole to examine you as it moves skittishly over the rocks.

Island kelpfish are favorite subjects of underwater photographers because they are generally quite easy to work with and they make striking photographic subjects.

SEA BASSES
(Family: Serranidae)

Kelp Bass: Often Called Calico Bass

Kelp bass (*Paralabrax clanthratus*) inhabit almost every kelp forest community in Southern California. The largest known specimen was over 28" long and weighed just under 15 pounds. With kelp bass, the larger they are the more wary they tend to be. Kelp bass are readily distinguishable by their bass-like shape, their markings, and their distinct coloration. Kelp bass are brown to olive-yellow above, yellow to off white underneath, and the body is usually mottled with white patches. They tend to hover in the kelp, and usually make subtle manuevers to keep some kelp between themselves and a diver, while being careful not to totally obstruct their view.

Black Seabass

Black seabass (*Stereolepis gigas*) are by far the largest fish seen in a reef community. Reaching a length of over 7 feet, a full grown black can weigh in excess of 500 pounds. Mature blacks tend to prefer rocky bottoms along the deep, outside edges of kelp beds, while the brick red juveniles are often seen over sandy bottoms. The juveniles have six irregular rows of black dots on their sides. These dots are evident in some adults when they are in a natural setting, but they quickly disappear when the fish is removed from the water.

For years black seabass were the most highly sought after prize by California sport fishermen, but heavy pressure from commercial fishermen reduced the population to the point of making black seabass an endangered species. It is no longer legal to take a black seabass, and seeing one underwater in its natural setting is a special thrill.

White Seabass

White seabass, *Cynoscion noblis*, are members of the croaker family (*Sciaenidae*), and their description can be found on page 130.

OCEAN WHITEFISH

(Family: Branchiostegidae)

Ocean Whitefish

Ocean whitefish (*Caulolatilus princeps*) are frequently encountered in Southern California, though they are rarely seen north of Monterey. Reaching lengths of up to 40 inches, ocean whitefish are yellow-brown above, and paler below. Yellow edging on the fins, especially the tail fin, is noticeable in many specimens.

Ocean whitefish are generally observed to be schooling along the outside edge of a reef. The schools often display curiosity about divers, and given time tend to come closer and closer.

Tending to cruise up in the water column just a few feet off the bottom, ocean whitefish prefer to feed upon plankton, squid, small fishes, and non-benthic crustaceans.

JACKS

(Family: Carangidae)

Jack Mackerel

Swirling schools of silver colored jack mackerel (*Trachurus symmetricus*) inhabit many reefs and kelp forests throughout the state. Individual fish are usually less than 10 inches long, but they are reported to reach lengths of 32 inches. Large schools contain thousands of fish, and they are truly a beautiful sight to see as the synchronized schools cruise the kelp on a sunny day. The presence of jack mackerel is often a tip-off that yellowtail and other predatory game fish are nearby.

Jack mackerel are usually observed to feed near the surface, but they are also known to probe into the sand for food.

SARGO OR GRUNTS
(Family: Pristipomatidae)

Sargo

Sargo (*Anisotremus davidsonii*) are similar in appearance to perch, but are actually members of the family that is commonly called grunts. The family obtains its common name from the sound that individuals make by grinding their well developed pharyngeal teeth together, a sound which is amplified by their swim bladder.

The bodies of sargo are silver with a dark vertical band behind the pectoral fin. In some specimens, additional less conspicuous dark bands are seen near the tail. Sargo are commonly seen in kelp beds and along rocky coasts. They tend to gather in loose schools and can be found from the shallows to depths of 130 feet. Visually it is difficult to distinguish sargo from perch, but it can be done. The anal fin of sargo contains 9 to 11 soft rays, while no perch has fewer than 13. If caught, sargo make grunting sounds when being removed from the water.

A favorite of sport fishermen and spearfishermen alike, thousands of sargo are taken annually. Sargo weigh up to 6 pounds and reach a length of 20 to 23 inches. While believed to be preyed upon by sea lions and some porpoises, sargo feed on small crustaceans, mollusks, and worms.

CROAKERS
(Family: Sciaenidae)

White Seabass

White seabass (*Cynoscion noblis*) are considered to be among the best tasting of all California fishes. They are really not true bass as their common name implies, but instead, are members of a family of fishes commonly referred to as croakers. Extremely wary fish, white seabass are almost never seen by Scuba divers. Most white seabass are seen only by expert free divers. Whites are usually seen along the deep, outside edges of the kelp.

Reaching lengths of up to 5 feet, a full grown white seabass will weigh close to 80 pounds. White seabass are bluish to gray above with dark speckling, and silver to white on the underside. Juveniles possess several dark vertical bars.

OPALEYE
(Family: Girellidae)

Opaleye

Opaleye (*Girella nigricans*) are in the sea chub family. As the name implies, the eyes are an opalescent blue. In addition, opaleye are characterized by their light gray-green to dark blue color, and single white spot on each side of the body. The largest opaleye ever recorded weighed 13½ pounds, and was 25⅜ inches long. Growth rings indicated that it was approximately 11 years old. Typical specimens weigh 3 to 4 pounds.

While reported from San Francisco to Magdelena Bay on the Baja Penninsula, it is only south of Point Conception that opaleye are observed year round. As adults, they are quite commonly observed near rocky bottoms, especially in areas where there are kelp beds. Inhabiting depths from 5 to 80 feet, they are most often seen in a range between 20 and 40 feet. Opaleye are quite active and commonly school.

Opaleye feed heavily upon algae and eel grass, but it is believed that they receive most of their nutritional benefit from the small animals which reside on the plants, rather than from the plants themselves. Opaleyes are fortunate, having no known specific predator.

HALFMOON
(Family: Scorpididae)

Halfmoon

Halfmoon (*Medialuna californiensis*) are also commonly called Catalina perch. They are dark blue above, becoming lighter blue on the sides as you look toward the bluish-white underbelly. Halfmoons reach a length of just over 1.5 feet, and weigh up to 4 pounds. Appearing in large schools, halfmoon are extremely abundant along the mainland and at the Channel Islands. Schools inhabit inshore reefs as well as kelp forests.

Halfmoon feed on a wide variety of sources including sponges, bryozoans, and red, green, and brown algaes. Like many fishes that inhabit the reef community, halfmoon are preyed upon by kelp bass, sand bass, large rockfish, and a variety of seals and sea lions.

SURFPERCHES
(Family: Embiotocidae)

Perch

Nineteen species of saltwater perch inhabit California waters. Some of the most noted species are the rubberlip surfperch (*Rhacochilus toxotes*), the black surfperch (*Embiotica jackson*), the walleye surfperch (*Hyperprosopon genteum*), and the barred surfperch (*Amphistichus argenteus*). Barred surfperch tend to inhabit regions with sand bottoms, while rubberlip surfperch and walleye surfperch prefer rocky reefs. Black surfperch can be distinguished from barred surfperch by the patch of enlarged scales found just below the pectoral fins of black surfperch.

As the name suggests, rubberlip surfperch feature rather prominent lips. The thick, fleshy lips are a whitish pink, and allow for easy, positive identification. Abundant in kelp beds, around jetties, and piers, and just outside the surf along open coasts, rubberlips are common from Mendocino County to Cape Colnett, Baja. A large rubberlip will reach a length of around 18 inches. They prey upon shrimp, amphipods, and small crustaceans.

Walleye surfperch are considered valuable by both commercial and sport fishermen. As you would likely suspect, they have very large eyes. These fish are steel blue on the upper body, becoming white on the lower surface. Walleye surfperch reach a maximum length of about 1 foot.

DAMSELFISHES
(Family: Pomacenntridae)

Blacksmith

Blacksmith fish (*Chromis punctipinnis*) are sometimes confused with halfmoon in southern waters, where both are common. Though their coloration is similar, blacksmith are much smaller. They rarely get as large as 12 inches, and most specimens are considerably shorter. Blacksmiths are blue-gray with numerous small black spots along the back, dorsal, and tail fin.

Blacksmith fish often form loose schools, that tighten either when the fish are being pursued or when they are being cleaned by senorita fish.

When being cleaned, the blacksmiths gather together in a tight ball up in the water column, and hang upside down waiting patiently for the senorita to do its work. If the senorita tries to quit before all the blacksmith fish are cleaned, the uncleaned blacksmiths will often try to block the senorita's path of escape. The cleaning behavior rapidly dissipates when divers get too close.

Garibaldi

Members of the family of damsel fishes, garibaldi (*Hypsypops rubicanda*) are among the most attractive of all California fishes. Especially attractive as juveniles, the reddish brown bodies of young garibaldi are decorated with brilliant blue, irridescent spots and stripes. The blue coloration usually disappears in the adult stage, though some specimens retain a faint blue border along their fins.

Garibaldi are protected by state law. However, they are not the state fish, or a state fish, as is often believed. The California state fish is the golden trout. Just the same, sport fishermen and sport divers should be aware it is illegal to catch, spear, or collect garibaldi for aquariums, without a commercial permit.

Garibaldi are quite common in the kelp communities of Southern California and are seen on almost every dive that is shallower than 100 feet. Garibaldi are not often seen at Santa Rosa Island, at San Miguel Island, or north of Pt. Conception. Reaching lengths of 15 inches, garibaldi seem to know they are protected, often displaying an intense curiousity about divers. If divers crack open a sea urchin, they are likely to be surrounded by a dozen or more swarming garibaldi within a matter of minutes.

The bright orange adults are highly territorial at times, and will vigorously defend their nests from intruders of any size during the spring. The nests are made from large patches of cultivated red algae. In making the nest, the male first selects the site, usually a flat face on a large boulder. The male then cleans the surface of all debris except for any of the desired red algae. Then the male leaves the nest to find more red algae which he brings back to the nest site. With other algaes removed, the red algae grows rapidly.

Then the male entices a female to lay her eggs in the nest. The eggs appear as small yellow balls in a sea of red. The male then fertilizes the eggs and vigorously defends the nest from all intruders until the young are hatched.

WRASSES
(Family: Labridae)

Sheephead

Sheephead (*Pimelometopon pulchrum*) are commonly observed residents of almost every rocky reef and kelp forest community throughout the Channel Islands and along the coast from Monterey south. As such, they simply cannot be omitted from any comprehensive discussion of California fishes. Sheephead are one of the larger members of the wrasse family, attaining lengths of up to 3 feet and weighing up to 36 pounds.

Sheephead are rather unusual fish, perhaps their most outstanding characteristic being their hermaphroditic quality. Born females, all sheephead change to males later in their lives. These wrasses begin their lives during the summer as eggs afloat in the pelagic environment. Upon hatching, the young fish spend several months at the mercy of open ocean currents. During this stage of their lives, the hatchlings are almost translucent, although some have several tiny black dots.

After a few months those juveniles that survive the perils of the open sea and that drift over a reef settle down to live in a reef community. For a few weeks the fish undergo dramatic changes in coloration. Within just a few days the sheephead become beautifully colored, being bright red or orange with a white laterally oriented line and black spots. Within a few weeks the vivid colors fade considerably, and the white lateral lines and dark spots disappear. The fish take on a reddish hue in their upper bodies, while the gut tends to be whitish.

After reaching the age of 4 or 5 years, sheephead achieve sexual maturity as females. Somewhere between the ages of 6 and 8, most become males. Identifying males and females is easy. In the male, both the head and the tail are deep blue to solid black. The lower jaw is white, and the mid body region is pinkish. Females are all red, though the shade will vary, and the heads of the females are much more rounded.

Sheephead feed heavily on sea urchins, some mollusks including abalone, some crustaceans, and some echinoderms. In fact, sheephead are often seen swimming about with a number of sea urchin spines stuck into their bony heads and gill covers. Sheephead have extremely hard heads, and are known to bump sea urchins off of rocks in order to get to their vulnerable tube feet. In the process, some spines occasionally become lodged in the bony head and gill covers. Large buck teeth enable sheephead to crush the shells of their prey.

Sheephead are often used by lobster fishermen as bait for traps.

Senorita Fish and Rock Wrasse

In addition to the sheephead, 2 other species of wrasse are often seen by snorkelers and sport divers in California. They are the very commonly observed senorita fish (*Oxyjulius californica*) and the rock wrasse (*Halichoeres semicinctus*). The female rock wrasse and both the male and female senorita have golden colored, cigar shaped bodies, but the senoritas lack the black spots found on the female rock wrasse. The male rock wrasse is grey with a black bar just behind the yellowish pectoral fin. Senoritas actively rid blacksmith fish, kelp bass, garibaldi, and other fishes of parasites.

As with many wrasses at night, both senoritas and rock wrasse bury themselves several inches deep in the sand. Icthyologists believe they do so for two reasons. First, the fish bury in order to hide from potential predators, and second, they use the coarse quality of the sand to knock parasites off of their own mucous coated bodies. The lubrication provided by the mucous is an adaptation that helps these fishes penetrate the sand.

WOLF EEL
(Family: Anarhuchadidae)

Wolf Eel

Wolf eels (*Anarrhichthys ocellatus*) are perhaps the fiercest looking of all the reef creatures encountered in California. Though capable of inflicting serious wounds with their canine like jaw teeth, their looks belie their true docile nature. Despite their physical appearance, wolf eels are not true eels, being much more closely related to blennies. The largest recorded wolf eel attained a length of just over 6 feet. Wolf eels feed upon market crabs and other small crustaceans, mollusks, and echinoderms. Though wolf eels inhabit many reefs, they are much more commonly seen the further north you dive. In fact, to many divers, wolf eels are synonymous with Northern California.

COMBTOOTH BLENNIES
(Family: Blenniidae)

Blennies

Blennies are small fish, most California specimens being less than 6 inches long. They have blunt heads, long, rather thin bodies, and are often beautifully colored in a variety of hues. Almost all species are found in close association with the bottom. Upon first being encountered, blennies often dart back into a hole, but their curiosity usually gets the best of them if you wait patiently for just a few minutes. Before long, they will come out to examine a diver that remains relatively still.

Three species of blennies are frequently observed by divers who take time to closely examine the cracks and crevices of California reefs. They are the rockpool blenny (*Hypsoblennius gentilis*) which is found south of Point Conception, the mussel blenny (*Hypsoblennius jenkinsi*) which can be seen south of Santa Barbara, and the bay blenny (*Hyproblennius gentilis*). Telling who is who within the blenny family is a difficult task to say the least. Even with a detailed scientific key, distinguishing one species from the next is hard to do, and in the field it is practically impossible for those who lack specific expertise.

GOBIES
(Family: Gobiidae)

Gobies

Although there are many types of gobies seen in California waters, three species of gobies are frequently seen in rocky terrain by sport divers. They are the Catalina or bluebanded goby (*Lythrypnus dalli*), the zebra goby (*Lythrypnus zebra*), and the blackeye or nickel eyed goby (*Coryphoterus nicholsii*). Bluebanded gobies are among the world's most striking fish. They have bright red-orange bodies that are covered with irridescent blue bars which match the color of their heads. These two-inch long fish are often seen in crevices with moray eels.

Though less common than bluebanded gobies, zebras have more stripes, which are also thinner. The blackeye goby is the largest of the three, reaching a length of 6 inches. Other than size, the blackeye goby can be distinguished by its whitish body and large black eyes.

The Sandy Plains

The
Sandy Plains

Throughout California waters there are great expanses of sandy bottom between the rocky reefs. Some of these plains extend for hundreds of square miles, while others are only a few yards wide. In any case, the marine life found in the sand is usually considerably different than that of a kelp forest, a rocky reef community, or the open ocean.

In many ways, these expanses of sand bear strong resemblance to deserts on land. During the day, the sandy plains can appear to be truly barren and boring, but at night, the same area can be alive with a variety of fish, octopi, crabs, clams, squid, sea pens, sea pansies, starfish, tube anemones, angel sharks, cusk eels, thornback rays, guitarfish, and more. Of course, there aren't always that many creatures to be found, but in many locations everywhere you look, you'll discover a pair of eyes staring back at you. And yet, at other times there is simply "no one home," and you can spend a lot of time just hoping to find some living thing to look at. In many ways, the sandy plains are like the open sea; it tends to be "feast or famine."

The most interesting sand based communities are usually in well defined areas where accumulated debris is decomposing and continuous supplies of phytoplankton are readily available. In the sand, there is a distinct lack of plant life, meaning grazers have little to eat other than decaying debris which has drifted into the area. In addition, sand dwellers do not have rocks for protection, so they must rely upon different adaptations to hide from predators and to disguise themselves from prey. Even the larger predators, like angel sharks and halibut, are very adept at changing their color pattern to blend in with the sand. In addition, like many sand dwellers, these fish often bury themselves, exposing little more than their eyes. Observant divers often find these creatures by seeing their outlines in the sand, rather than by seeing the entire body or even body parts.

138

Most species of sand dwellers have extremely low profiles. A major difference between the sand and crevice-filled reef habitat is that other than in or on the sand itself, in sandy terrain there are few places to hide. In essence, most animals that reside in the sand must be 1) excellent burrowers, 2) be able to rebury themselves rapidly after they are exposed, 3) be able to stabilize the substrate around them in some way, or 4) be superbly camouflaged if they are to successfully survive. As examples, many species of clams and crabs are excellent diggers. Clams spend a considerable amount of time buried in the sand, with only their paired siphons exposed. The siphons enable clams to take in oxygenated water, to feed, and to eliminate wastes without exposing themselves to danger. Creatures like tube anemones and sea pens appear to be little more than uninteresting plants if uncovered during the day, but at night the feeding mechanisms of these filter feeders become more evident. If uncovered, these species can quickly rebury. In many sites groups of sand dollars are so thick that the association works to stabilize the sand, and the same is true for several species of worms. Flatfish such as turbot, halibut, sole, and sanddabs use their splotchy coloration to camouflage themselves in the sand.

Flatfish are able to closely match both the color and the pattern of the sand around them. Experiments have shown that the fish must be able to see their surroundings, and then they are able to alter the shape of special black colored pigments in the cells of their skin so the fish can closely resemble their surroundings. The specialized skin cells are called iridocytes.

Moving away from a beach, as the water becomes deeper and the effect of surge is reduced, a greater variety of sand dwellers becomes evident. Starfish, white sea urchins, turbot, several species of sculpin, cabezon, and a number of snails are commonly found. Brittle stars are seen by observant snorkelers and divers who detect the exposed tips of their snake-like looking arms. The central discs of brittle stars are often completely buried. Horn sharks, too, are at times found resting on the sand bottom. And if the sand patch lies downcurrent from a nearby kelp bed, any number of additional species can be expected. At greater depths, water motion from wave action is reduced and some grazers can be found feeding on a fine layer of decaying matter.

California Sand Dwellers

Cnidaria

Sea pansies
Sea pens
Tube anemones

Arthropoda

Sheep crab
Spider crab

Ectoprocta

Bryozoans

Mollusca

Boring clam
Octopus
Squid
Snails

Echinodermata

Sand dollars
Sea stars (starfish)
Sea urchins

Chordata

Cartilaginous Fishes

Electric rays
Thornback rays
Bat rays
Shovel-nose guitarfish
Skates
Angel sharks

Bony Fishes

Spotted cusk-eels
Flatfish
  Halibut
  Sole
  Turbot
  Flounder
  Sanddab
Sculpin
Sarcastic Fringeheads

The Sand Invertebrates

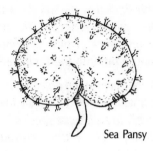

Sea Pansy

Fleshy Sea Pen

Phylum: Cnidaria
Sea Pansies

Sea pansies (*Renilla kollikeri*) are a form of colonial anthozoan, being closely related to sea pens. Members of the phylum Cnidaria, sea pansies appear as purple, heart-shaped discs embedded in the bottom. Sea pansies use a stalk on the underside to anchor in the sand. Modified polyps called siphonozoids are found on the disc. These tiny eight armed structures serve in creating a water current which circulates through the colony. The current supplies a renewed source of oxygenated water that is utilized by the sea pansy. Food is taken in through another group of modified polyps called autozooids which are also located on the disc.

Sea pansies utilize another interesting adaptation to secure food. The colony secretes a net of mucous which is supported by the polyps. When the net becomes full of entrapped food, the net is swallowed.

Sea pansies are brilliantly luminescent, and like sea pens, are noted for their light emitting capacity. They are commonly seen in sand from depths of 20 feet and deeper.

Sea pens

Colonial animals, all species of sea pens are described in the phylum Cnidaria. Sea pens generally require quiet waters and are not often found close to the surf zone. They are supported by a vertically oriented, central, limy axis made from calcereous secretions, which is bordered by polyp bearing lobes. The lobes are attached to a bulb which is buried in the sand. During the day and in surgy conditions the lobes retract, and

141

Bioluminescence

If you have ever been swimming, diving, or boating in the open sea at night, you were probably thrilled by the way the water around you seemed to glow. The lighting effect is the result of a phenomenon called bioluminescence, which is almost always witnessed by those who venture out into the Pacific after sunset.

Bioluminescence describes the ability of an organism to emit light. This phenomenon occurs in a wide variety of marine invertebrates and in some vertebrates as well. In fact, some species in over 50% of all animal phyla display this rather curious ability. Bioluminescence is the result of a chemical reaction in which chemical energy is converted to radiant energy, or light. The reactions are extremely efficient, with almost no heat being produced as a by-product. In fact, in scientific circles bioluminescent displays are often referred to as cold biological light. Some, but not all types, of bioluminescence involve a variety of forms of bacteria. The chemical reactions are varied among the different species that display bioluminescent properties, and most of the chemistry is rather well understood by the scientific community.

At various times, many areas in California waters are filled with single-celled bioluminescent planktonic dinoflagellates. At night, these animals glow when mechanically stimulated from wave action, or from the kick of a fin or swipe of a hand. When the dinoflagellates are present in dense concentrations, their flickers of light are bright enough to outline a swimmer or diver.

Bioluminescent displays by sea pens and sea pansies in California waters can be stimulated by night divers by lightly touching the colony of animals. The touch will often generate a wave of light that travels down the organism.

In many species of deep water fishes and squid, bioluminescence has a distinct survival value, being utilized to frighten or confuse predators, and to attract prey. Some fishes such as midshipman use their bioluminescent quality to attract their mate. The role in lower forms of invertebrates is not well understood, though it is generally believed to have evolved eons ago in an effort to prevent certain types of oxygen poisoning.

the stalk slides into the sand leaving only a couple of inches exposed. Sea pens are noted for their light producing capacity, or bioluminescent property. At night noticeable flashes of light are sometimes emitted when the polyps are disturbed.

The three most common species seen in California are the sea pen (*Acanthopilum gracile*), the pink or fleshy sea pen (*Leioptilus guerneyi*), and the white or slender sea pen (*Stylatula elongata*). Pink sea pens are the most visually outstanding, with the soft white axis bearing fluffy pink lateral lobes which create a feather-like appearance. Generally found in deep water with a coarse sand substrate, pink sea pens are seen throughout the state. Larger specimens stand 2 feet above the sand bottom and are common in Central California and Northern California. As its name suggests, the slender sea pen is distinguished by its tall, thin design reaching a height of more than one foot but being only 2 to 4 inches wide.

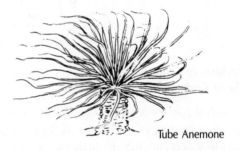

Tube Anemone

Tube Anemones

When actively feeding, tube anemones, often called cerianthus anemones, are considered to be some of the most striking of the sand based invertebrates. Their scientific description is *Cerianthusa estuari*. Tube anemones construct long, mucous-lined tunnels into which they retract when the animals are inactive or disturbed. The tubes are laced with active stinging nettles, cleverly deposited by the anemone.

When feeding, two rings of tentacles are extended into the water column. The longer tentacles of the periphery, which typically reach 4 to 5 inches beyond the tube, are used to capture fish and other prey, and the shorter tentacles around the mouth are most useful for feeding purposes. Like the tube, the tentacles are colored a light brown, but their flower-like appearance and numerous extended tentacles create an excellent photographic subject out of these rather blandly colored creatures.

However, some tube anemones appear to be a bright pink underwater, but when photographed with an artificial light source, the anemones appear a drab brown. The apparent change in color is due to the intensity of the strobe lights that are normally used by underwater photographers. At depth when sunlight has been naturally filtered, the tube anemones radiate a different wavelength of light than they do when lit from close range by a powerful strobe.

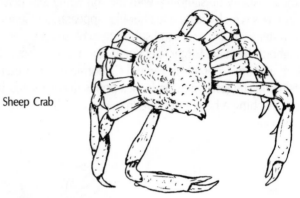

Sheep Crab

Phylum: Arthropoda
Sheep Crab and Spider Crabs

Several species of crabs that are commonly called spider crabs or sheep crabs frequent sandy plains. The two most notable of these species are the sheep crab (*Loxorhyncus grandis*) and the spider crab (*Loxorhyncus crispatus*). Both sheep crabs and spider crabs make dramatic photographic subjects, appearing much like a miniature version of what one might imagine a prehistoric sea monster to look like. They have pear shaped bodies which taper to two prominent points at the leading edge of the carapace. The carapace is the portion of the exoskeleton which covers the cephalothorax of various arthropods.

Sheep crabs attain a larger maximum size, attaining a carapace width of 6 to 7 inches, while spider crabs are usually less than 2 inches across the carapace. Sheep crabs possess formidable claws, but when approached, they tend to give ground. Their long legs provide startling agility. Spider crabs are often referred to as decorator crabs, a name they acquire from the behavior of taking small anemones, sponges, rocks, shells, algae, and other debris and attaching some or all to their own shell. While the carapace of spider crabs is almost always covered by debris, sheep crabs often host parasitic barnacles.

Lacy Bryozoan

Phylum: Ectoprocta
Bryozoans

In many sites where the bottom is composed of sand and clay so that the substrate is not constantly shifting, divers encounter a number of orange colored masses that look much like a lace-like imitation of the petal of a rose. The 2 to 6-inch wide flower-like structure is actually an upright colony of animals called bryozoans. They are not well studied, but it is known that most are hermaphroditic, meaning that each animal possesses both male and female reproductive organs. Fertilization occurs when eggs and sperm join after being released into the water. Two commonly observed species of bryozoans are the fluted (*Hippodiplosia insculpta*) which has a curly leaf-like shape, and the lacy (*Phidolopora pacifica*) which has a definite lace-like look. Both species are filter feeders, preying on plankton and organic decay.

Wart-necked
Piddock Clam

Phylum: Mollusca
Boring Clam

California waters are inhabited by several species of clams collectively called boring clams. A commonly seen species in sand or mud communities is the wart-necked piddock, *Chalae ovidae*. Like other clams, wart-necked piddocks have an incurrent and excurrent siphon through which they gather food and excrete wastes. The rest of the animal lives in a shell that is burrowed in the substrate. Wart-necked piddocks dig their burrows by rotating their rasp-like shell in order to cut a hole out of the bottom.

When observing these clams, you are likely to see only the purple colored tips of the siphons which protrude from the bottom when the animal is feeding. Clams are amazingly fast burrowers, especially when disturbed. If you try to dig a clam out of the sand or mud, you are unlikely to ever be able to even see the shell if you do not dig with true intensity. Upon first sighting a clam, many divers have no idea that the organism they are looking at is even an animal, much less a clam, as the siphons have a plant-like appearance.

Octopus

Octopi are cephalopods. They are mollusks which lack an external shell, but their soft body parts are quite similar to hard shelled mollusks. In addition, the foot has been transformed into tentacles equipped with suction cups. A number of species that are commonly seen in the sand are classified only by their genus, Octopus.

Octopi are extremely clever and well adapted for their life in the sea. (The word octopuses and octopi are both correct plural forms of the word octopus.) The most commonly observed of the sand dwelling species of octopi is Octopus rubencens. These small octopi are almost always less than one foot from tentacle tip to tentacle tip. They are primarily nocturnal, and are rarely seen out in the open during the day. At night they emerge from the protective confines of their dens in order to hunt. Octopi are often discovered resting on the sand just outside the mouth of their dens. Normally their coloration is very similar to the surrounding sand, though when disturbed they can change their coloration, shape, and the texture of their skin in dramatic fashion. An octopus can be dark brown and smooth one moment, cherry red and ruffled the next, and white and smooth the next. The range of possibilities seems limitless, as the different modes are believed, at least in some instances to indicate anger, fear, and frustration etc. Light altering cells in the skin, called chromatophores, enable octopi to vary their color, while muscular contractions allow the changes in texture. When changing color, octopi do not emit light. Instead they expand or contract the chromatophores which alters the shape of the pigment granules that are contained within the chromatophores. The change in the granules causes a change in coloration.

The unique ability to change color and shape so rapidly allows octopi to be superb camouflage artists. Many times while filming an octopus I have lost the animal out in the open sand. As I started to swim away, my diving buddy has pointed the animal out to me. To my astonishment, the octopus was in the exact same place it was throughout my filming efforts, only it had changed color, shape, and texture to blend in with the sand.

The sex life of octopi is rather interesting. In some species the male places sperm on a modified tentacle which is then offered to the female. When the female accepts his offer, she takes both the sperm packet and the majority of the male's tentacle as well. The female then fertilizes herself with the packet. The male is capable of regenerating his lost arm. The female lays the eggs inside of her den where she guards them for the next two months until they hatch. Shortly afterwards the female dies.

In California waters, octopi feed upon a variety of creatures, their favorite food sources being lobster, a variety of clams, shrimps, and scallops. Several species of octopi readily prey upon tuna crabs (*Pleuronocodes planipes*) when they are available. While up in the water column, tiny kelp scallops sometimes attach themselves to the tuna crabs, and as the scallops grow, the crabs become too heavy to support themselves and sink to the bottom. The octopus sits very still and waits for a careless, overweighted crab to stray too close. As the crab draws near, the octopus extends a tentacle. When the crab gets within an inch or two, the octopus quickly reaches out with the tentacle and envelops the crab. Then the octopus pulls the entangled crab under its mantle and bites it. I have seen octopi that were only 10 inches across capture and devour as many as 5 tuna crabs in 20 minutes.

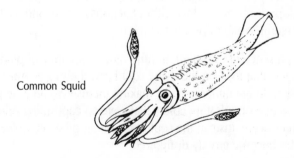

Common Squid

Squid

Squid (*Loligo opalescens*) are usually found in, or over, deep water but they mate and lay their eggs in much shallower surroundings. Of all the phenomena that occur in the sand, squid spawns are perhaps the most fascinating. During heavy "runs" which can last for several weeks, literally millions of squid gather to mate in coastal canyons and along steep sand drop offs at the offshore islands. After successful fertilization the females proceed to lay their eggs, and then the adults die. The eggs are laid inside of 4 to 8-inch long, white egg casings which are firmly attached to the sandy bottom. Each egg case contains close to 200 squid, but it is estimated that on the average only one of the 200 will become an adult, as the hatchlings are heavily preyed upon by a variety of carnivorous fishes and crustaceans. In many places the casings are so thick you literally can not see the bottom. The once light brown sandy substrate takes on the appearance of a luxurious, creamy white shag carpet for as far as you can see. In 5 to 7 days you can see the bright red eyes of the yet to be hatched squid by holding the egg between a light source and your eye. In another week or so the eggs will hatch, and the newborns will instinctively head for deep water.

If in any given year the squid do spawn in depths accessible to sport divers, it is generally during the middle of winter. But that is not always the case as heavy spawning has also been documented in August near San Diego. As the squid, driven by strong instincts, search frantically for partners, they pay little attention to the presence of outsiders, whether predators or diving photographers. It is believed that the squid live for only one year, and their strong desire to mate is the way nature has insured the perpetuation of their species.

Squid are capable of swimming forward, backward, and sideways with equal rapidity, propelling themselves with a directable siphon and their undulating tail fins as they seek out a mate. The actual act of mating occurs most often during nocturnal hours. When in the act, these cephalopods change colors within fractions of a second as they pulsate from creamy white to purple to brown.

The presence of the squid, in turn, attracts an array of predators and scavengers that feed on the dying squid and their eggs. After mating, the squid deteriorate rather quickly, making themselves an easy prey. The kaleidoscope of pulsating colors that were so captivating prior to mating no longer occur. Instead the squid take on an off-white color and their tentacles become grossly disfigured.

Along the egg covered sea floor, a diversity of creatures feed on the dying adults and the eggs. Horn sharks and angel sharks eat so many squid they simply can not force down another bite. These predators rest on the bottom with partially eaten squid dangling from their mouths, as if immobilized by their consumption. Rockfish, black seabass, and white seabass often join the feast. Lobster and crabs leave the protective

148

crevices of the reef to forage on both the dying adults and the eggs. But even with all this activity, the squid die off so fast that in places they are stacked one on top of the other in piles that are up to 2 feet high.

High up in the water column the feast continues for bat rays, sea lions, pilot whales, and blue sharks as they prey upon live squid. Blue sharks rarely come so close to shore, but when a natural food source presents itself, these open ocean sharks readily come into shallow water to feed. At times the blue sharks are present in such large numbers that everywhere you look, you will sight a blue shark swimming, mouth agape, through the squid.

It is fascinating to observe the blue sharks as they feed on the thick concentrations of squid. The squid make easy targets, and apparently the sharks have no mechanism to tell them that they are full. The blues eat and eat until their stomachs are totally distended, and there are squid hanging out of their mouths. At that point the blues begin to regurgitate. When they have created some additional room, once again they begin to munch their way through the concentrations of squid. Such a scene tends to make us think of sharks as proverbial "eating machines." But remember: In the wilderness, there are no guarantees or free meals. When an opportunity presents itself, those animals that intend to survive must take maximum advantage of existing situations.

Lewis' Moon Snail

Snails

Many species of snails are known to reside in the sand. One of the most common species is Lewis' moon snail (*Polinices lewisii*). These snails attain a diameter of close to 5 inches. Lewis' moon snails have well developed olfactory senses that help them locate their favorite foods — clams and other snails. Lewis' moon snails are characterized by their large gray foot and gray mantle that almost covers its shell. The shell is brown, gray, or off-white and has a very large aperture.

When traveling across the sand, snails almost always leave a distinct, grooved track which is quite easy to follow.

The Submarine Canyons

Adjacent to some plains, the geography changes dramatically as large submarine cayons dominate the marine topography close to shore in many areas of the state. By name, some of the more prominent submarine canyons are the La Jolla Canyon, Scripps Canyon, Redondo Canyon, Dume Canyon, Mugu Canyon, Monterey Canyon, Carmel Canyon, Delgada Canyon, Mattole Canyon, Eel Canyon, Mendocino Canyon, and the Sur-Partington Canyon. These canyons do not form a continuous trough.

On a worldwide scale the origins of submarine canyons are rather diverse. However, like the majority of submarine canyons, the California canyons typically have V-shaped profiles with high steep walls, some terrace-like ledges, numerous rock outcroppings, a winding course, a number of tributaries, and a constantly shifting substrate. The topography of individual canyons changes dramatically from gentle slopes to sheer cliffs that descend vertically into the depths. In between the two extremes there are a variety of slopes and cascading ledges constructed in stair step-like fashion. The composition of the sediment in the submarine canyons of California varies from primarily mud and silt combined with organic decay, to fine grained quartz, to rock.

The sediment layers in the canyons are usually quite unstable due to erosion from water motion, organic erosion, and land slides that constantly change the face of the terrain. Strong currents and deep water upwellings are common in many canyons. Their presence is evidenced by erosion on the canyon walls and by the constant freshening of nutrients in canyon waters. Many canyon dwellers such as anemones, boring clams, sea pens, sea pansies, rays, and flatfish are constantly digging up the bottom as they burrow, and this action leads to further erosion. Erosion caused by living organisms is known as organic erosion.

Similar to deep water upwellings, coastal currents often help enrich the canyon areas by supplying a constant source of nutrients. Their flow is often blocked by canyon walls, a factor which leads to additional erosion and which also serves to trap nutrients. In those nutrient rich areas, impressive quantities of lower invertebrates are often found, establishing the foundation for many canyon food chains.

Almost every time divers find rocks or any type of solid bottom on the sand, they are more likely to discover a greater variety of life. Where the sand is constantly shifting, life can be more difficult for those who want to take up residence. In flat areas, especially those just below shifting sands, an abundance of life is often present. Distinct thermoclines are often encountered in the canyons, and noted differences in the marine life found above and below the thermocline are quite evident.

Several of the canyons are noteworthy due to their massive size or their popularity with divers and fishermen. Perhaps the most notable are the La Jolla Canyon and Scripps Canyon off the coast of La Jolla in San Diego and the Monterey Canyon which heads off Moss Landing near Monterey. The La Jolla Canyon which extends to approximately 4/5 of a mile south of the pier at Scripps Institute of Oceanography is a favorite night diving site for many San Diegans. Likewise, the Monterey Canyon provides exceptional diving for divers in the Monterey area. The Monterey Canyon is the largest and deepest submarine canyon along the west coast of the United States. The 3-branched canyon is over 51 miles long and it maintains its steep walls and canyon-like look at a depth of 9,600 feet before the floor flattens out. Geologists often compare the canyon to Arizona's Grand Canyon.

As is the case with the just mentioned canyons, most submarine canyons border deep vertical basins. The upper lip of the shallowest canyon walls are, therefore, in close proximity to deep water. The end result is that some animals which normally live in deep waters are apt to visit the upper reaches of the canyons more often than other nearby sites. This seems to be especially true at night, adding a lot of excitement to night time exploration.

151

Phylum: Echinodermata

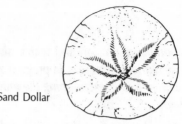

Sand Dollar

Sand Dollars

Sand dollars (*Dendraster excentricus*) are echinoderms, sharing many common characteristics with sea urchins. The spineless, disc-shaped bodies of dead specimens are familiar to almost everyone who has ever shown an interest in the marine environment. The five petal, flower-like design is a distinguishing feature.

Sand dollars are considered to be colonial animals. That is because a sand dollar is composed of many individual polyps that share a common skeletal case, or test. Sand dollars are typically 3 to 4 inches in diameter.

Orienting with the prevailing current, sand dollars frequently occur in numbers so large that their half-buried bodies almost obscure the bottom. Common from the intertidal zone to about 50 feet, the presence of so many sand dollars helps prevent the sand from shifting, thus stabilizing the substrate. Detritivores, sand dollars feed primarily upon decaying organic matter. Food particles are entrapped in a sticky mucous that is among the spines. A solitary pink barnacle (*Balanus tintinnabulum*) is found atop many specimens.

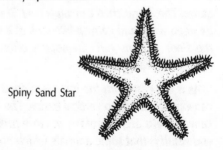

Spiny Sand Star

Sea Stars (Starfish)

Several species of sea stars can be oberved on the sandy terrain where they live. Among the most numerous are southern starfish, *Astroecten brasiliensis armatus*, which are sometimes called spiny sand stars or simply sand stars. This species attains a diameter of up to 10 inches. Almost always a grayish color, these sea stars can be easily recognized by the rows of spines along the margins of the arms.

The sand star (*Petalaster foliolata*) can be distinguished from the southern star by its thicker arms, longer spines, and slightly smaller width. This species is also quite common.

It is interesting to observe the feeding habits of many species of sand dwelling sea stars. These animals often find their prey buried deep in the sand. To get to its desired quarry, the sea star extends its stomach far down into the burrows dug by its prey. The sea star envelops its catch with its stomach, and then the digestive process begins.

Heart Urchin

Sea Urchins

From Santa Barbara south, in deeper, cooler waters, divers occasionally discover clumps of small white sea urchins. This species is called the white sea urchin (*Lytechinus anamesus*). While their bodies can take on an orangish hue, these urchins are often snow white. Sometimes white sea urchins are spread out and solitary, but when scavenging on a common source they will clump together in softball-sized groups. Individuals attain a maximum diameter of about 2 inches.

Another interesting sand dweller is the heart urchin (*Lovenia cordiformis*), sometimes called the mouse urchin. Heart urchins look like a spine covered ball of fur, with the longest spines generally curved back over the body. This species often buries itself in the sand, exposing only the long, pliable spines. When uncovered, mouse urchins can crawl across the bottom at a rapid pace. Once settled, the urchins tend to quickly rebury themselves. When handling mouse urchins, it is amazingly easy to get punctured by a spine. While the wound is usually not serious, it can be irritating.

The Sand Vertebrates
Phylum: Chordata

Cartilaginous Fishes: Rays, Skates, and Sharks

Electric Rays

Electric rays (*Torpedo californica*) are sometimes found hovering over sand patches. An electric ray is characterized by its flat blue-gray upper body, light underbelly, enlarged pectoral fins, and short tail that has laterally flattened fins which serve as stabilizers. Unlike other rays, electric rays have highly developed electrical organs that are composed of muscle tissue in which the normal electrical generating capacity has been dramatically increased. The organs, one on each side of the rounded body disc, are filled with over 375 stacks of cells that work much like the electrical plates in batteries. Four large nerve trunks enable the rays to have voluntary control over the electrical organs, allowing the animals to discharge an electrical charge as either a defensive response or to stun prey. Though the rays cannot maintain a constant current, the electricity they generate is very similar to that found in your home. Worldwide there are approximately 30 species of electric rays, and the electricity they generate varies considerably from species to species. One non-local species has been documented to produce up to 220 volts at 20 amps.

The shock of the electric rays that inhabit California is not strong enough to seriously hurt a swimmer or diver, but if you are hit, you will definitely feel an unpleasant jolt. Electric rays are reported to feed regularly on fish, crustaceans, and mollusks, and have been observed while feeding upon dying squid and squid eggs during squid spawns.

Thornback Rays

Thornback rays (*Platyrhinoidis triseriata*) feature several rows of small spines that extend down the midline of their backs. These rays attain an overall body length of 2 to 3.5 feet. Thornback rays have brown backs and white or cream colored underbellies. Ranging from San Francisco to Baja, these rays are common residents of many sand communities. They are generally seen resting on the sand and often turn away when approached by divers. Thornback rays feed primarily upon small crustaceans, mollusks, some echinoderms, and a variety of fishes.

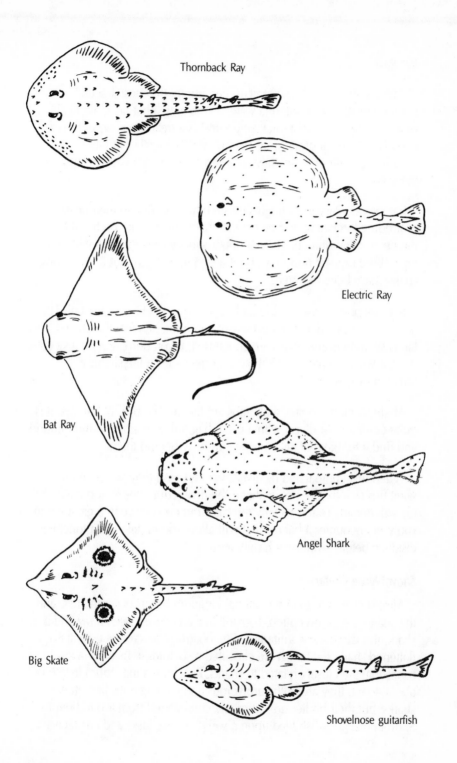

Thornback Ray

Electric Ray

Bat Ray

Angel Shark

Big Skate

Shovelnose guitarfish

155

Bat Rays

Bat rays (*Holorhinus californicus*) are among the most graceful of marine creatures in California waters. Watching a bat ray majestically weave its way through a thick kelp forest is a treat divers never tire of. Usually solitary, at other times bat rays gather together in large squadrons. In either case, they are often encountered while resting in the sand.

Bat rays are easy to distinguish from other rays due to their fleshy, protruding foreheads, and by the fact that their eyes are on the sides of the head, rather than on the flat upper surface as is the case with most rays. These rays are normally about 4 to 5 feet across, but 8 foot "wing spans" have been reported.

Bat rays possess formidable barbs located at the base of the tail. The barb is used as a defense mechanism when the rays feel threatened. As far as swimmers and divers are concerned, if you don't step on a sleeping bat ray, you are highly unlikely to experience even the slightest threatening display.

Abalone, clams, oyster, and crabs are the favorite prey of bat rays. It is quite common for divers with a catch bag full of abalone to turn around and find a hungry bat ray cruising curiously behind them.

When first approached by divers, bat rays tend to raise up on their wing tips as if to give notice that if the diver continues to approach, the ray will depart. That is usually the case, but every once in a great while you will encounter a bat ray that will allow you to get within touching distance before swimming hastily away.

Shovel-Nose Guitarfish

Almost every group in the animal kingdom has its eccentricities, and the sharks are no exception. Looking like a cross between a ray and a shark, the shovel-nose guitarfish (*Rhinobatus productus*) is essentially a flattened shark. Guitarfish are classified as batoids, which are a group of fishes that are closely related to sharks. Reaching a maximum length of about 4 feet, they are both carnivorous and cartilaginous like other sharks, but their bodies are extremely compressed from top to bottom. Shovel-nose guitarfish feed upon a variety of mollusks and crustaceans.

156

Not much is known about their life cycle, but these batoids are obviously well adapted for their life on the bottom. They breathe through spiracles rather than gills. Water is drawn in through the spiracles which are on the top of the head. The water then passes over the gill tissues before being expelled through the gill clefts on the underside. The spiracles help bottom dwellers adapt to life in the sand, in that it is very likely that delicate gill tissues would be severely damaged by coarse sediment if they inhaled water while resting on the bottom. By breathing through the valved spiracle, guitarfish protect their vital and rather sensitive gill tissue.

Skates

Skates are cartilaginous fishes. Upon first glance, they strongly resemble rays, but there are at least 3 significant biological distinctions. First, skates lack the stingers or barbs found at the base of the tail as is the case in many rays. Second, the tails of skates are lobed, fleshier, and heavier than those of rays. And third, the broader tail of skates often contains many spines, although using this characteristic alone will not provide positive identification. The largest skates are much smaller than even medium sized rays. Many skates have electrical tissue in their tails, but the output of electricity is minimal.

There are at least 8 species of skates that have been reported in California waters. Of these, the big skate (*Raja binoculata*) is certainly the most commonly seen by swimmers and divers. These skates are found in depths that range from 10 feet to 360 feet from San Quintin Bay, Baja, to the Bering Sea off the coast of Alaska. While known to attain a length of 8 feet, few specimens grow longer than 6 feet. Like most bottom dwelling sharks, rays, and skates, big skates feed on a wide variety of mollusks, crustaceans, echinoderms, and small fishes. Big skates are preyed upon by some bottom dwelling sharks.

Angel Sharks

Angel sharks (*Squatina californica*) are characterized by their unusually flattened body, terminal mouth, and greatly enlarged pectoral fins. They are common residents in many sandy environments. Their normal coloration varies from sandy gray to reddish brown, with a white underbelly and dark spots above. Angel sharks attain a maximum length of 5 feet and weigh up to 60 pounds. They are generally observed resting on the bottom, either partially or completely buried. Because of their dorsoventrally compressed bodies (meaning flattened from top to bottom), the unthreatening appearance of their face, and their docile nature, angel sharks are often mistaken for skates or rays.

Angel sharks feed upon a variety of small fish, crustaceans, and mollusks. They do not readily come to bait, and are known to lie in a rather dormant state for extended periods of time. When frightened by divers, angel sharks flee along the bottom. Often as one angel shark swims away, another previously unseen angel shark will also bolt, and then another and another, in a domino effect until as many as 20 or more are gliding gracefully along the bottom.

In recent years an angel shark fishery has been developed along the California coast. This commercial venture is quite worrisome to many scientists who have only recently discovered that angel sharks are much more territorial than was once thought, and they are not nearly as numerous as many fishermen seem to believe.

Bony Fishes

Spotted Cusk-eel

Spotted Cusk-eels

Spotted cusk eels (*Otophidium taylori*) are small eels, reaching a maximum length of close to 15 inches. Ranging from northern Oregon to San Cristobal Bay in Baja, they are usually only encountered at night on sandy bottoms. These eels are easy to identify due to their barbels or chin whiskers, and the dark spots along their thin, almost translucent bodies. At night they hunt octopus, small fish, and a variety of crustaceans. When threatened or alarmed they can bury themselves in the sand by rapidly wiggling their way in, tail first. During the day, spotted cusk-eels are rarely seen, remaining buried in the sand, several inches below the surface.

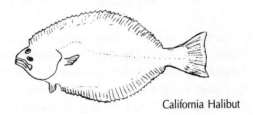

California Halibut

Flatfish

Halibut

The California halibut (*Paralichthys californicus*) is included in a group of fishes commonly referred to as flatfish. This group includes sole, turbot, flounder, sand dabs, and halibut. As adults, all of these species have both eyes on one side of the head.

The California halibut is easily distinguished from other flatfish in California by noting 3 characteristics. California halibut possess a large mouth, many sharp teeth, and a high arch in the lateral line above their pectoral fin. They are usually a speckled brown on top and white underneath, although all brown and all white halibut are known.

159

California halibut are by far the largest of the flatfish found in state waters. The species ranges from Oregon to Magdelena Bay, about halfway down the Baja Peninsula, but specimens show up north of Morro Bay only on a seasonal basis.

Of all the fish species that can be found in a sandy environment, the California halibut is the most heavily sought after by fishermen. Both experienced spearfishermen and seasoned rod and reel enthusiasts are well aware that halibut are usually found in sand areas in less than 60 feet of water. Sport fishermen troll with large anchovies when possible, but halibut will hit a variety of live bait, dead bait, and artificial lures.

Halibut are slow growers, as five year old fish average only 15 inches in length, but large female halibut do reach sizes of up to 50 inches and 50 pounds. Halibut feed upon anchovies and other similarly sized fish, and are preyed upon by a variety of sharks, electric rays, sea lions, seals, and some inshore dolphins.

Other Flatfish

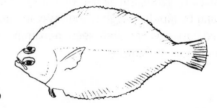

Pacific Sanddab

In addition to the California halibut, several other species of flatfish inhabit state waters. The most prominent are the C-O turbot (*Pleuronichthys coenosus*), the diamond turbot (*Hypsopsetta guttulata*), the curlfin turbot (*Pleuronichthys decurrens*), the Pacific sanddab (*Citharichthys sordidus*), the speckled sanddab (*Citharichthys stigmaeus*), the rock sole (*Lepidopsetta bilineata*), and the starry flounder (*Platichthys stellatus*), all of which are sand dwellers.

All flatfishes are born with one eye on either side of the head, and as they mature one eye migrates over so that as adults both eyes are located on one side of the head. Halibut and sanddabs are in the family scientifically named Bothidae, a family commonly called lefteye flounders. The common name is derived from the fact that in juveniles the eyes are located on either side of the head, but as the fish mature, the eye on the right side migrates to the left side, at least in the majority of specimens. So as adults both eyes are most often found on the left side.

California
Sandy
Plains

1. A fleshy sea pen is composed of a colony of tiny polyps.

2. An octopus seeks safety inside an empty shell.

3. Several species of octopi are commonly seen at night along the sandy ocean floor away from the safety of their dens.

PLATE C-1

4. Squid mating above egg casings.

5. Camouflaged curlfin turbot.

6. Sculpin.

7. Red crabs overweighted by attached kelp scalle make easy prey for predators.

8. Angel sharks inhabit sandy plains near reef communities.

PLATE C-2

Spotted cusk-eel.

10. Sheep crab.

. Sand dollar.

12. California halibut.

3. Red crabs are small, often numerous, but are not commercially sought after.

PLATE C-3

14. Bat star.

15. Sand star.

16. White sea urchins prey upon a red sea urchin.

17. Sand rose anemone.

18. Rays often bury themselves in sand exposing only their eyes and spiracles which are used in respiration

PLATE C-4

19. Thornback ray.

20. Bat ray.

21. Tube anemone.

22. Horn shark.

23. Horn sharks are so named for the modified dermal denticles atop the dorsal fin. Here a diver examines a juvenile.

PLATE C-5

California
Open
Ocean

24. Blue shark.

25. Diver hand feeding a blue shark.

26. Sleek and elegant, blue sharks are the most commonly observed open ocean sharks in California.

PLATE C-6

7. Compared to blue sharks, mako sharks have a stockier build. Makos have larger gill slits, a more pointed snout, thick caudal keel, and a large symmetrical tail.

8. Ocean sunfish can often be seen near the surface in the open sea.

PLATE C-7

29. Close-up of pelagic red crab.

30. Scalloped hammerhead shark.

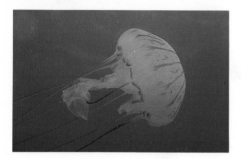

31. California purple striped jellyfish.

32. Close-up of jellyfish tentacles.

33. Usually observed in the open sea far from land, pilot whales occasionally visit shallow inshore waters

PLATE C-8

There are 22 species of righteye flounders (family Pleuronectidae) found in California water including rock sole, starry flounder, C-O turbot, and diamond turbot. As adults, both eyes are most often found on the right side of the head of righteye flounders. As with other flatfish, their coloration patterns vary, and their eyes can be found on either side of the head as adults. Identification is tricky without a detailed pictorial reference source.

Sculpin

Sculpin

If you ask a group of Scuba divers if they saw a sculpin after any one dive in California, odds are the answer would be "yes, in fact we saw several." You see, the term sculpin is a common name used to describe many different kinds of fish in two separate families. While it is certainly easier for sport divers to speak in laymen's terms, the widely used word, sculpin, is a classic illustration of why scientists must refer to fish by a specific genus and species.

The family Scorpaenidae, the scorpionfishes, contains 62 species, the most species of common California fish found in any one family. Included in this grouping is the common sculpin (*Scorpaena guttata*), sometimes called the spotted sculpin. All rockfish are also found in the Scorpaenidae family, in the genus *Sebastes*.

However, 56 additional species which inhabit California waters are also commonly called some type of sculpin. These fish make up a family named Cottidae. Confusing? To well schooled icthyologists, no, but to fishermen, divers, and other recreational enthusiasts it certainly can be.

At least 10 species in the Cottidae family are frequently observed. Those species include sailfin sculpin (*Nautichthys oculofasciatus*), snubnose sculpin (*Orthonopias triacis*), buffalo sculpin (*Enophrys bison*), tidepool sculpin (*Oligocottus maculosus*), and grunt sculpin (*Rhamphocottus richardsoni*), and cabezon (*Scorpaenichthys marmoratus*).

Some of these species are found in both sandy and rocky substrates, while others show definite preferences for one habitat or another. The tiny grunt sculpin are seen much more often in Northern California than in the south. Cabezon are known to prey heavily upon abalone populations, but cabezon do not usually occur in large numbers. These fish are a favorite of many fishermen, but you should be aware that the roe can be quite poisonous. Cabezon are characterized by their large, bizarrely shaped heads, fan-like pectorals, stumpy tails, goggle eyes, and tufted "eyebrows." The bodies of the females are a mottled gray-green, while the males are a splotchy, reddish brown.

Sarcastic Fringehead

Sarcastic Fringehead

One of the most bizarre looking fish that inhabits the sandy environment has an equally unusual name. It is the sarcastic fringehead (*Neoclinus blanchardi*). Reaching a length of about 12 inches, sarcastic fringeheads are found in crevices and holes in sandy or muddy bottoms. Their bulbous eyes, heavily frilled faces, and deep purple to brown coloration make positive identification relatively easy. Sarcastic fringehead are members of the Clinidae family, the same family that various kelpfish belong to.

Like octopi, one spot blennies, and small crabs, sarcastic fringeheads are often found living in discarded cans and bottles.

Life in the Open Sea

Life in the Open Sea

With the exception of the beaches, environmental conditions in mid-ocean change more dramatically and faster than in other sectors in the sea. In many ways, life in the open sea can be described as "here today and gone tomorrow." In somewhat of a feast or famine setting, the presence of many animals is difficult to predict. Vitally important open ocean life forms vary from tiny planktonic plants and animals which occupy the lowest level of many food chains to the large apex predators. While the apex predators occupy the top rung of their respective food chains, their very existence is predicated upon many microscopic sized organisms. Most boaters and divers rarely as much as notice the presence of planktonic life except when dense concentrations drastically alter the water color and reduce visibility. But plankton plays a major role in the food chain of almost every animal in the sea.

The exact whereabouts of planktonic concentrations are controlled by many factors, one of the most significant being currents. The currents, in turn, are significantly affected by several variable forces including wind, tide, swell, and temperature. As a result, water movement is quite difficult to accurately predict, and in turn, so is the presence of concentrations of planktonic life forms. Significant concentrations may occupy only an acre, or can cover thousands of square miles.

Most forms of plankton are quite sensitive to sunlight, and while they live close to the surface of the ocean, they often descend during daylight hours to avoid sunlight, and ascend for feeding at night. Wind, currents, and the presence or lack of a thermocline can also have great bearing upon the location of plankton concentrations.

In taking a closer look at the food chain in the open sea, you will learn that there are two broad classifications of plankton. They are 1) phytoplankton which are classified as plants, and 2) zooplankton, which are animal forms. Of the phytoplanktonic forms, diatoms and dynoflagellates are the most abundant. The term zooplankton describes a range of animals varying from the larvae of what will later develop into benthic animals to permanent planktonic forms such as small crustaceans. Examples of these larval forms include scallops and lobsters. Zooplankton graze upon phytoplankton.

The presence of both phytoplankton and zooplankton attracts larger animals like sardines and anchovies. These small pelagic fishes are preyed upon by species such as mackerel and top smelt which attract yellowtail, albacore, and barracuda. These middle predators in turn are sought after by pelagic sharks and marine mammals. It is, therefore, rather easy to understand how the subsistence of animals at the top of the food chain is so dependent upon planktonic life.

Other invertebrate species such as jellyfish, salp, comb jellies, and by-the-wind-sailors also appear in enormous quantities at times, only to be almost totally absent at others. Life in the open sea is truly "here today and gone tomorrow," as the migratory species constantly pursue those forms that drift with currents.

The Open Sea Creatures

Cnidaria
By-the-wind sailors
Jellyfish

Ctenophora
Comb jellies

Chordata

Cartilaginous Fishes	Bony Fishes
Blue sharks	Albacore
Mako sharks	Bonito
Scalloped hammerheads	Mackerel
Thresher sharks	Swordfish
Basking sharks	Striped marlin
Megamouth sharks	Yellowtail

The Seamounts

Seamounts are undersea pinnacles that rise dramatically from the deep ocean floor, creating sheer vertical walls, ledges, plateaus, caves, cracks, and crevices that are ideal living quarters for a great many marine animals. The pinnacles lie unprotected in the open sea and are constantly awash with currents and nutrient rich upwellings, creating an extremely prolific environment. There are several prominent seamounts in California waters that can be visited by both fishermen and sport divers, and most are associated with the Channel Islands. While not likely to discover life forms that would not be seen elsewhere, the diversity and quantity of life is often staggering.

The seamounts are rocky reefs, though in many places this fact is hard to believe upon first sighting. The rocks are often totally obscured for as far as you can see by a shag carpet-like covering of invertebrates. Large mats of aggregate anemones (*Anthopleura elegantissima*), dense populations of bright red, orange, and pink *corynactus* anemones, and large white Metridiums are present almost everywhere. The thick stalks of the Metridiums are often more than 12" long, making great photographic subjects.

Nestled in amongst the anemones are chestnut cowries, sponges, barnacles, colonies of encrusting bryozoans, several species of well camouflaged crabs, worms, dinner plate sized rock scallops, and a host of colorful sea stars. For those with a creative eye, navanaxes, nudibranchs, and in many places purple coral create the enviable problem of what to photograph first.

Some crevices are filled with large lobsters, morays, cleaner shrimp, and wolf eels. Others harbor a variety of rockfish. Many people believe California fishes, other than the garibaldi, are rather drably colored, but few notions are farther from the truth. Some of the more striking examples of California fishes include vermillon, rosy, starry, copper, black and yellow, and gopher rockfish as well as treefish, blue-banded gobies, sculpin, and ling cod. At the seamounts divers always have a chance to look around the corner and see a giant black seabass. Huge blacks weigh up to 500 pounds, and sometimes travel in pairs. Being eyeball to eyeball with a giant seabass is sure to make any dive a memorable event. Protected by law, black seabass cannot be legally taken by fishermen due to their endangered status.

Schools of mackerel, barracuda, and yellowtail are common sights at the seamounts. Ocean sunfish often hover around the pinnacles. Ocean sunfish are often referred to by laymen by their scientific name, *Mola mola*. These rather bizarrely shaped fish are always fun to encounter. Some days divers can swim right up to them, but if you do so, you are likely to be sorry. They have very fine scales on their skin, and their skin is coated with a thick layer of mucous that adheres to anything and everything it touches, especially divers. On other days, the first exhaust from Scuba bubbles sends ocean sunfish hurriedly on their way. Many times divers encounter a beautiful California purple striped jellyfish or a salp chain floating past in a mid-ocean current. And yes, blue sharks constantly cruise many seamounts, being attracted by the presence of so many sources of food.

Open Sea Invertebrates

Phylum: Cnideria

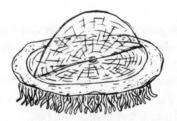

By-the-wind Sailor

By-The-Wind Sailors

Occasionally great expanses of the surface of the open sea are populated by literally thousands of by-the-wind sailors (*Velella lata*), which are also referred to as purple sailing jellyfish. Large gatherings cover several acres — the result of population density and prevailing winds and currents. These jellyfish reach diameters of about 4 inches, with the clear bell being surrounded by a blue to purple border. A distinguishable feature of by-the-wind sailors is the small sail which sits atop the bell. On days when there is very little wind and the surface is flat, putting your eyes as low to the surface as you can, and then looking out into a dense concentration of by-the-wind sailors reveals a truly remarkable sight.

California Purple Striped Jellyfish

California Purple Striped Jellyfish

Worldwide there are more than 200 species of jellyfishes, and while several species are seen in state waters, the most common and most striking is the California purple striped jellyfish (*Pelagia panopyra*). These animals are relatively primitive, having only a simple system of nerve nets, and lacking both a brain and elaborately controlled muscles. The nerve net near the surface of the bell controls the rhythmic pulsations used in swimming, and another more diffuse net controls reactions and feeding.

Their bodies are hollow with a digestive cavity in the center, and being rather poor swimmers they tend to go wherever water motion takes them. California purple striped jellyfish have white to milky white bells with regular longitudinal brown or brownish-purple stripes. They grow to slightly more than 1 foot in diameter and the tentacles typically trail up to a length of 5 to 6 feet behind the bell.

These animals are remarkable predators due to their use of small stinging cells called cnidoblasts, which are possessed only by members of the phylum Cnidaria. Cnidoblasts are located both in the tentacles, where often more than 100 can be found near the surface, and around the edge of the mouth. The cninoblasts of various species are laced with toxins intended to paralyze the jellyfish's prey. The cninoblasts are capable of penetrating human skin, and the reaction to the stings will vary considerably.

California purple striped jellyfish prey upon plankton and small fish. Juvenile croakers and other small fish which are either immune to or protected from the stinging cnidoblasts can be found hiding amongst the tentacles of some jellyfish, while small crabs are sometimes seen living on the bell.

168

Phylum: Ctenophora

Comb Jellies

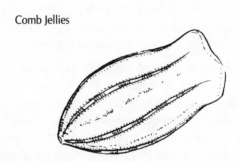

Comb Jellies

Comb jellies are members of the phylum Ctenophora. They look like jellyfish at first glance, but upon closer observation you can easily distinguish between them. Comb jellies take on a variety of shapes, unlike jellyfish which are generally bowl-shaped.

The single most important distinguishing feature between comb jellies and jellyfish is that comb jellies have 8 ciliated "combs," which appear as lines covered by tiny hairs on the body. Though capable of independent locomotion via the wavelike movement of the cilia, comb jellies usually go wherever the current and wind takes them. For that reason, you will usually see hundreds or none on a given day, but it is unusual to see only a few. Many forms are bioluminescent, and in only one non-California species do the tentacles contain stinging cells.

Oceanic Food Chains

Plants and animals can be classified in many ways. As examples, animals can be described as scavengers or predators, as vertebrates or invertebrates, as herbivores or carnivores, or as primarily active during the day or at night. Plants might be categorized as either true flowering plants or as algaes. Plants and animals can also be described by their trophic relationships to one another. Trophic association involves an analysis of what plants and animals eat and what eats them. All food chains are predicated upon trophic associations.

For example, the terms herbivore and carnivore describe trophic relationships. Herbivores are animals that eat plants while carnivores prey upon other animals. Describing plants and animals by their trophic relationships is all encompassing, as all plants and animals require food and energy for survival, growth, and reproduction.

Marine organisms are generally said to be (1) producers, (2) consumers, or (3) decomposers. Plants are the primary producers. Plants are considered to be self nourishing, meaning they use only solar energy in the photosynthetic process in order to create organic matter. As the primary producers, plants are placed in the first trophic level. Animals that feed on plants are placed in the second trophic level, while carnivorous animals (animals that prey upon other animals) occupy the third trophic level. Animals that prey upon the carnivores of the third trophic level are considered to occupy the fourth trophic level. All animals are considered to be consumers, while the decomposers consist primarily of bacteria and fungi.

It is rather important at this point to distinguish between the flow of essential nutrients and the flow of energy in any given ecosystem. The movement of nutrients and gases is circular, going from plants, to animals, to decomposing bacteria, and back to plants. Energy transfer, on the other hand, is unidirectional, going from the sun to plants to animals. Extensive laboratory and field studies demonstrate that the transfer of energy from one trophic level to the next ranges only between 6% and 20%, with 10% being considered about average. The important point to be aware of is:

It takes a lot of plants to support a smaller number of small animals, and it takes a lot of these smaller animals to support an even lesser number of larger predators, and so on. The inefficient transfer of energy explains why there are so many more small marine animals than large ones. It also helps explain why many large predators are solitary hunters, in that it takes fewer smaller animals to support a single larger predator. The larger predators need not school for safety, because they have fewer natural predators. Most large sharks for example are solitary animals (although there are exceptions).

The paths through which the transfer of energy takes place in nature is called a food chain. Marine food chains are generally described as (1) grazing chains or (2) detritus chains. Grazing chains begin with plants and lead through a series of immediately related grazers and predators. Detritus chains are based on the waste and death of members of the grazing chains. Detritivores are animals that feed upon decaying organic matter. It is worthwhile to note that many marine animals, such as lobster eat other animals and also feed on organic decay, thus playing vital roles in both chains simultaneously.

Very few straight food chains exist in nature. In fact the term "food web" more accurately describes the most common interrelationships in which many animals that occupy a fourth trophic level in some webs, can occupy the fifth or sixth level in others. Sharks are a good example. A shark that feeds on anchovies fits into the fourth level. But if the shark is feeding on a yellowtail that ate a mackerel that ate the anchovie, then the shark occupies the sixth trophic level. However, that shark might be eaten by another shark and that would represent still another trophic level in that association.

The main point is that oceanic food webs are highly interconnected. That means that when any portion of one web is altered the ramifications are far reaching, having an impact throughout the web. Removal of a given species due to an abusive fishery, or what might appear to be localized problems due to pollution, often has far reaching effects throughout the web in areas that can be thousands of miles away.

Open Sea Vertebrates

Phylum: Chordata

Sharks

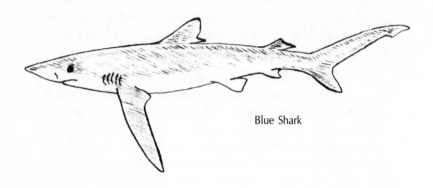

Blue Shark

Blue Sharks

So named for their beautiful coloration, blue sharks (*Prionace glauca*) are the most common of the open ocean sharks of California, although precise population dynamics on many species are difficult to obtain. Blue sharks, members of the requeim family, are easy to recognize due to their royal blue color, their long, slender build, and their comparatively long pectoral fins. Blues inhabit temperate seas all over the world. Typically they range in size from 5 to 8 feet, and at 8 feet weigh in the neighborhood of 70 to 80 pounds. There are unverified reports of a 13 foot long blue in California waters, and I have on many occasions filmed blues that were at least 10, and perhaps 11 feet long.

The beautiful blue color is most obvious on sunny days when the sharks are near the surface. Their irridescent skin sparkles as shimmering rays of sunlight reflect off the sharks' backs. Like many open sea predators, blue sharks are well countershaded, being more lightly colored on the underside. When looking up toward a blue that is near the surface, the whitish underbelly tends to blend in with the lighter water colors found near the surface. And when looking down onto a blue from above, their darker topside tends to blend well with the blue black waters below.

Extremely graceful swimmers, blues tend to glide rather effortlessly at a speed only slightly faster than most humans can easily swim. They rarely seem to swim at "full throttle," but make no mistake about it, when they want to, blues can move in a hurry. In fact, they are believed to be among the fastest of the sharks. Like most of the world's more than 350 species of sharks, blues use their long, powerful tails for thrust, controlling their attitude and turns with their pectoral fins. Highly manueverable, blues give the impression of always being in total control, normally operating at a level well below their maximum capabilities.

As with most sharks, blues tend to be sexually segregated. In Southern California, we see all males in the summer and all females in the winter. The two sexes are believed to mix for brief periods of time during both spring and fall. The males can be easily identified by noting the pair of claspers on the underbelly, and the females by the lack of the same. Claspers are part of the males' reproductive system, and a pair of claspers is possessed by the males of all cartilaginous fishes. Exactly where and how they mate is still a matter of scientific speculation. Females are often scarred with the imprints of the teeth of males. The scars are believed to be the result of the males biting the females during the act of mating, and they are referred to as mating scars. The females benefit from an interesting morphological adaptation in that the skin covering their gut, pectoral fins, and head is up to 3 times thicker than that of the males. This adaptation is found in many shark species, and it is believed to prevent the females from being harmed while mating.

Blues are often accompanied by small fish known as common remoras (*Remora remora*) which attach themselves to the body of the shark. Their behavior is not completely understood, and while it is clear that the remoras benefit from the presence of the sharks, scientists are not sure if the sharks benefit, are harmed, or remain unaffected. Remoras feed on scraps of food dropped by the host shark, and the remoras also benefit by the mere presence of the shark. That presence serves as protection for the remora from other potential predators who tend to give way to the shark making life a little safer for the remora. Remoras also obtain food by cleaning parasites off their hosts, and at times leave the host to capture food. Although blue sharks are not known to feed on the remoras, remoras have occasionally been found in the stomachs of blue sharks.

Like many sharks, when blues open their jaws to feed, a thick, white membrane covers their eyes as a means of protecting them from the bites of competing sharks and from the bones of fish upon which they often prey. The protective eyelid is called a nictitating membrane, and it raises up to cover the sharks' eyes. The movement of the nictitating membrane sometimes creates the optical illusion that the sharks' eyes roll back in their head when they eat.

The diet of blue sharks consists primarily of squid and small schooling fishes such as anchovies and sardines, though blues are also considered to be opportunistic feeders. While working for the National Geographic Society on a prime time television documentary about sharks, Howard Hall and I filmed blues feeding on squid (*Loligo opalescens*) at night on the backside of Catalina Island. During heavy runs, when the squid are mating, there are literally millions of squid within a few square miles. The heavy concentration of squid brings in the blue sharks, as well as sea lions, seals, bat rays, pilot whales, dolphins, giant seabass, and more.

On nights with no moon, squid are attracted to the powerful lights of the commercial squid fishing boats. Swimming below the boats outside of a shark cage amongst the sharks and squid, Howard and I repeatedly filmed blue sharks gorging themselves with squid. The blues ate and ate until their stomachs were totally distended and their mouths completely full. At that point they would begin to regurgitate squid. As soon as they had created some room, the blues once again began swimming mouths agape as they munched their way through dense concentrations of their prey.

While blue sharks are considered by the scientific community to be opportunistic feeders, they are certainly not indiscriminate about what they eat. When chumming for blues, it is imperative to use a bait source that is part of the sharks' natural diet. As with other shark species, fishermen have found some rather bizarre items in the stomachs of blue sharks. Blues may end up with odd items in their stomachs at least in part as a result of the design of their mouth and teeth. Like many marine predators, blue sharks have the ability to extend their upper teeth well forward when feeding. When in the act of closing their mouths, the teeth fold back towards the stomach, making it rather difficult for their natural prey, and any other odds and ends which might have gotten sampled, to get away. These sharks feed by tearing and swallowing. Blues do not chew their food as humans do. It is well documented that there is an extremely high concentration of acid in their stomachs which greatly aides the digestive process. The "extendable" jaw and digestive capablitity are adaptations that help blue sharks and many other fishes be successful predators.

Mako Sharks

Shortfin Mako Shark

Another of the migratory species of sharks that is commonly seen in California waters during late spring, summer, and fall is the shortfin mako shark (*Isurus oxyrhincus*), sometimes called the bonito shark. Makos are a part of the family Lamnidae, which consists of a few fast swimming species, including great white sharks. In photographs and scientific descriptions, makos are easily distinguished from blue sharks by 1) their conically shaped snouts, 2) thick, yet fusiform bodies, 3) extremely large gills which allow for highly efficient gas exchange, 4) their pronounced caudal peduncles which forms the caudal keels, and 5) their homocercal, or almost perfectly lunate-shaped tail fin.

In the water, telling the two species apart is even easier. Simply put, blues are beautiful, makos are frightening. A mako looks like a gaggle of teeth attached to the front end of a torpedo. Rows of long, narrow teeth are almost always exposed, giving the impression that they are too large to fit in the shark's mouth. When excited or nervous, makos move in a herky-jerky fashion, darting back and forth in a rather unnerving manner. This is a vivid contrast to the effortless, graceful appearance of blues. Numerous long, thin parasitic copepods are often attached to the dorsal fin and are also evident in the mouth of makos, where they cause nasty looking scars. The combined effect of rows of exposed teeth, the wounds, and the streaming parasites give many makos a look that Hollywood producers would love to put on the face of any sea monster.

Despite their fearsome impression, makos are extremely careful about what they bite. While blues will swim right up to a bait, makos are more leery, and by the time they get up the courage to approach, they tend to be very excited. In the wild, makos feed upon a variety of fast moving fishes such as albacore, swordfish, and other sharks. Makos are truly built for power and speed as they must be in order to capture such fast moving prey. Fishermen are well aware that makos tend to follow fast moving albacore populations. When frustrated fishermen catch only the head of an albacore, the culprit is often a contented mako.

In fact almost everything about a mako has a look that says "I am built for power and speed." Mako sharks have comparatively large gill slits which helps enable these predators to metabolize large amounts of oxygen in a matter of seconds. Makos also have a high percentage of red muscle fiber. Red meat generally has a higher rate of blood flow, meaning more oxygen can be supplied to the muscles in a shorter period of time. Body temperatures which are constantly maintained at a level a few degrees higher than ambient temperature allows muscles to respond faster and stamina is increased as well. The lunate tail and the thickness of the caudal peduncle, sometimes called the caudal keel, help generate speed as well. The caudal peduncle is a short ridge located just in front of the tail near the midline of the body which gives the tail additional support.

As menacing as makos do appear, on many occasions I have witnessed sea lions harassing makos, biting them repeatedly in successful efforts to chase them out of an area that was full of bait that our film crew had placed into the water. The sea lions return to the scene and feed to their heart's content, as if the whole encounter was simply "no big deal."

Scalloped Hammerheads

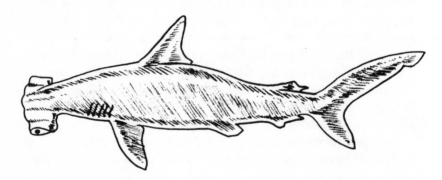

Scalloped Hammerhead

Scalloped hammerheads (*Sphyrna lewini*) are occasionally seen in Southern California during summer and fall, especially when surface temperatures are higher than normal. Obviously, hammerheads are characterized by their flat, wide, unusual, hammer-like shaped heads. Scalloped hammerheads can be distinguished from other hammerheads by noting the pronounced ridges which create a scalloped effect along the leading edge of their heads. When fully grown, scalloped hammerheads usually attain a length of close to 10 feet.

In recent years, this species of shark has been the subject of many television documentaries and magazine articles due to their now famous grouping or schooling behavior in Mexico's Sea of Cortez and other tropical seas. During certain times of the year, divers regularly see gatherings of up to 400 or more scalloped hammerheads in these areas — but in California waters, these sharks tend to be solitary.

It is very unusual for apex predators, animals at the top of the food chain, to school. Lacking significant numbers of natural predators, there is no reason to think that scalloped hammerheads school due to the concept of "safety in numbers." It is likely that mating or migratory behavior is at the heart of the issue, but no one is really sure.

Like many species of pelagic sharks, when in the open sea scalloped hammerheads feed primarily on smaller fishes, squid, and pelagic rays. When hunting in reef communities, scalloped hammerheads are known to feed on a variety of fishes, crustaceans, and mollusks.

Other Sharks

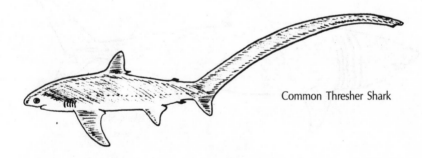

Common Thresher Shark

Thresher sharks, Basking sharks, Megamouth

While several other species of pelagic sharks are known to inhabit California waters, four species are especially interesting. They are the bigeye thresher (*Alopias superciliosus*), the common thresher (*Alopias vulpinus*), the basking shark (*Cetorhinus maximus*), and megamouth (*Megachasma pelagios*).

Like all threshers, both the bigeye and the common thresher are characterized by their enormously long tail fins, which can be half as long as their bodies. Bigeyes, which are rare compared to common threshers, can be further characterized by the facts that (1) they only have about 10 or 11 teeth on each side of the upper jaw, while common threshers have 21 or 22, (2) the dorsal fin of bigeyes is located well back on the body almost even with the pelvic fins, and (3) bigeyes, as their name suggests, do possess very large, upward looking eyes. The large eyes are indicative of the fact that bigeyes are deep water sharks, though they do come to shallower waters when feeding at night. Like white sharks and makos, threshers maintain a body temperature that is higher than surrounding water temperature. This adaptation allows them to live and prey successfully in colder, deep waters. Threshers are known to use their extremely long tails when feeding in order to herd small schooling fishes such as sardines, anchovies, and herring towards their own mouths. Threshers are also known to prey upon squid, small tuna, hake, and other fishes. Bigeye threshers normally range no farther north than San Clemente Island, while common threshers range from central Baja to Canada.

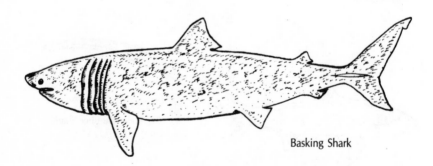

Basking Shark

Of all true fishes, basking sharks (*Cetorhinus maximus*) are second in size only to whale sharks. Basking sharks reach a length of 45 feet long, though most individuals attain a length of only 30 feet. A 30 foot long specimen that was caught near Monterey weighed 8,600 pounds. Like the whale shark, basking sharks are filter feeders. Basking sharks swim mouth agape through dense concentrations of plankton, and continuously take in enormous quantities of water which is strained for food by their huge gill rakers.

The bodies of basking sharks are characterized by being grayish-black above, shading to paler colors below, and their skin is studded with densely packed thorn-like dermal denticles. Basking sharks are rarely seen by Scuba divers, but they are commonly seen by boaters in deep waters off the coast of Central and Northern California. The range of basking sharks extends from the Sea of Cortez to Alaska.

The name megamouth describes a large, deep water shark whose very existence had been questioned until recent years. In 1984, a megamouth was caught in a net off of Catalina Island. Verification confirmed that both a species and a family of sharks that were thought to be extinct for centuries is, in fact, still alive. Megamouths are presumed to be quite large. Plankton feeders, their mouths are located in the terminal position at the front of their heads. Over 100 rows of tiny teeth are located in both the upper and lower jaws. Obviously little is known about the habits of megamouth.

Other Fishes

Albacore

Schools of fast moving albacore tuna (*Thunnus alalunga*) frequent state waters in summer and fall. "Albies," as they are popularly called by fishermen, are enthusiastically sought after by sportfishermen and commercial fishermen alike. Albacore are highly migratory, ranging from Clarion Island off the Pacific coast of mainland Mexico all the way to Alaska. They tend to prefer clear, blue water within temperate seas. The largest documented albacore weighed 93 pounds and was 5 feet long. Albacore feed on a variety of food sources including squid, pelagic red crabs, some forms of plankton, and a variety of small fishes. In turn, albacore are preyed upon by swordfish, marlin, and a number of species of sharks.

Bonito

Schools of bonito (*Sarda chiliensis*) frequent California water in late spring, summer, and fall. Reaching a length of up to 40 inches, bonito are dark blue above with silverish underbellies. Bonito range from Chile to the Gulf of Alaska, and are usually encountered only well out to sea. Members of the tuna family, bonito are fast swimmers.

Mackerel

Several species of mackerel play vital roles in open ocean food chains. These fishes include Pacific mackerel (*Scomber japonicus*), bullet mackerel (*Auxis rochei*), and skipjack (Euthynnus pelamis). Pacific mackerel attain sizes of up to 25" and 6.3 pounds. Bullet mackerel are slightly smaller, reaching a maximum size of 20" long and 5 lbs., while skipjack can be up to 40" long and weigh as much as 35 lbs. These fishes are members of the family named Scombridae, as are bonito, albacore, and other fishes commonly called tuna. Pacific mackerel, bullet mackerel, and skipjack feed mostly upon small fishes and shrimp-like creatures called krill. These fishes are also known to feed opportunistically on other bite sized animals, especially juvenile squid. Porpoises, sea lions, yellowtail, marlin, sharks, black seabass, and other large predators prey upon small mackerel and skipjack.

Striped Marlin

The round, elongated bill, triangular dorsal fin, and movable pectoral fins will distinguish striped marlin (*Tetrapturus audax*) from all other fish in state waters. Striped marlin generally appear in Southern California in summer and fall, ranging north to Point Conception. Most specimens weigh less than 250 pounds, but the largest known individual reached a length of 10½ feet, and weighed 465 pounds.

These marlin prefer to feed upon pelagic fishes such as sardines, jack mackerel, bonito, and flying fish. They are also known to hit squid and crabs.

Swordfish

Swordfish (*Xiphias gladius*) inhabit warm temperate seas all over the world. In California waters they are most plentiful south of Point Conception, and are most commonly observed from June through September. However, when the water is uncommonly warm such as it often is during intense El Nino Currents (see page 32), swordfish have been known to travel as far north as Oregon. Swordfish are occasionally seen leaping out of the water, and have also been filmed as deep as 2,000 feet by camera systems that were mounted on submersibles. When seen at the surface, swordfish can be recognized by their large sickle-shaped dorsal fin. The tail fin generally breaks the surface as well.

The largest documented swordfish was 14 feet 11¼ inches long and weighed 1,182 pounds. Swordfish can be distinguished from other fishes

Yellowtail

Albacore

Bonito

Pacific Mackerel

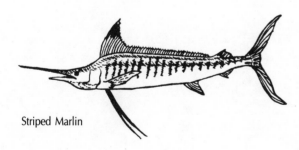

Striped Marlin

Swordfish

by 1) their flattened sword which forms the greatly extended upper jaw, and 2) their lack of both scales and pelvic fins. Squid are a favorite food source of swordfish.

Yellowtail

Ranging from Cabo San Lucas at the southern tip of Baja to as far north in some years as Monterey Bay, yellowtail (*Seriola dorsalis*) are a popular game fish in California water. Yellows are usually schooling fish that are found close to shore, near offshore islands, and over offshore banks, but they are also attracted by mid-ocean kelp paddies. Yellowtails have been documented to attain lengths of just over 5 feet. The west coast record is listed at 80 pounds, and a 111 pounder was caught off New Zealand. Except for albacore during the short lived height of the season, yellowtail are the most highly sought after game fish in the state.

Yellowtail feed primarily during daylight hours. Their favorite food sources are considered to be pelagic red crabs, anchovies, squid, sardines, and small mackerel, though they are also opportunistic feeders and will often take whatever food is available.

Interestingly, yellowtail are often seen rubbing up against the tails of blue sharks. Scientists debate whether the yellowtail are ridding themselves, the sharks, or perhaps both, of parasites.

California
Marine
Mammals

California
Marine Mammals

From San Diego to the Oregon border, California waters are enriched with a diverse population of marine mammals. At various times of the year state waters are frequented by numerous species of whales, dolphins, porpoises, seals, sea lions, and sea otters. One or more of these species of mammals inhabits all of the previously discussed habitats. Sea lions and seals can be sometimes observed resting on rocky shores, in kelp forests, over sand, and even out in the open sea. Whales and dolphins are most often seen in the open ocean, while sea otters are commonly observed in kelp forests and hunting in reef communities.

I do not think anyone fully understands mankind's affinity for mammals, but undoubtedly they do hold a special place in our view of the animal kingdom. Perhaps we feel some common bond, a kindred spirit with creatures who are so similar to ourselves. And too, as a species we probably feel some guilt and sorrow for the cruelties we have wrought on so many marine mammal populations. In any case, being in the wild with marine mammals represents a treasured occasion for all of us, and few places provide a better opportunity than California.

As a result of mankind's intense interest and because mammals frequent so many habitats, I have decided to include a special segment devoted strictly to California's marine mammal populations. And before delving into individual species, a brief overview of the characteristics of mammals is in order. Marine mammals are the descendants of creatures that once lived on land. As these mammals evolved, they developed specialized adaptations which equipped them for survival in the aquatic environment. Like their land dwelling relatives, marine mammals breathe air and nurse their young. Some display body hair, and several species possess the skeletal remnants of legs which were utilized by their predecessors eons ago.

All marine mammals are presented with a problem by the fact that they often spend extended periods of time in water. Water conducts body heat away much faster than does air, and even in the temperate waters of California, the problem of keeping warm is one that must constantly be fought. Whales combat the situation by insulating their bodies with a thick layer of fat, called blubber. Seals and sea lions use a fat layer as well as body hair, or fur, to provide thermal insulation. The fat layers also provide a source of energy for these creatures when other food sources are not available. Sea otters lack a fat layer and are forced to rely on their dense fur and the energy obtained from their voracious feeding habits.

Common Marine Mammals of California

Cetaceans

Baleen Whales

Gray whales
Blue whales
Finback whales
Minke whales
Humpback whales

Toothed Whales

Pilot whales
Common dolphin
Bottlenose dolphin
Killer whales

Pinnipeds

Sea lions

California sea lions
Stellar sea lions

Seals

Harbor seals
Elephant seals
Northern fur seals
Guadalupe fur seals

Mustelidae
(Weasel Family)

Sea otters

Cetaceans

All whales, dolphins, and porpoises are classified as belonging to the order Cetacea. The order has been subdivided into 2 suborders; Mysticeti, the baleen whales, and Odontoceti, the toothed whales. Baleen whales lack teeth, and capture their food by filtering the water. Utilizing two rows of tough, flexible, horny sheets of modified hair called baleen to strain the water, the whales swim open mouthed through dense concentrations of krill and plankton. The baleen looks like a synthetic fiber, and a mouth full of the "fibers" looks a great deal like an oversized scrub brush. When the whales close their mouths, the water is expelled and the food is trapped by the baleen. Baleen whales are also known to feed upon small fishes.

The toothed whales include all dolphins and porpoises. Whether to use the term dolphin or porpoise poses a problem for many people. In casual conversation the words are interchangeable. For more exact usage, the term dolphin refers to long-beaked whales, while porpoise describes those with smaller, stubby noses. Toothed whales feed primarily upon squid, octopi, fish, and in some cases such as with killer whales, they feed on other mammals.

Baleen Whales

California Gray Whales

Of all the marine creatures that can be observed in California water, few have captured the public's attention like California gray whales (*Eschrictius robustus*). It is relatively easy to observe gray whales in the wild along the Pacific coast of western North America during their southward migration which occurs during late fall and early winter. On days of high concentrations, as many as 75 whales will pass given coastal points. As a result, the variety of whale watching expeditions offered to the public are well worth joining.

Gray whales are so named because of their mottled coloration, a combination of natural black skin and large patches of white barnacles which attach themselves to the whales in significant numbers. Fully mature specimens reach lengths of nearly 50 feet, and weigh more than 40 tons. As large as those numbers might seem, grays are relatively small whales. By comparison, blue whales which are also baleen whales, reach lengths of over 100 feet and weigh more than 100 tons.

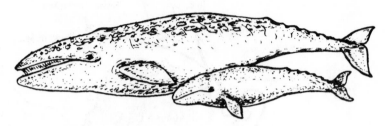

California Gray Whale and Calf

The life cycle of California gray whales centers around their feeding grounds in the Bering Sea and their calving and breeding grounds in the lagoons of Mexico's Baja Peninsula. At a speed of 3 to 5 miles an hour, almost every mature California gray whale swims 10,000 miles annually during a round trip migration. Even more astonishing is the fact that many experts believe these whales cover the entire distance without any significant feeding. The trip represents the longest mammal migration known to science. The whales spend the late spring, summer, and early fall of the northern hemisphere feeding upon plankton and amphipods in prolific northern waters. In late fall, they begin their journey south. Groups of southbound whales number as high as 15, though smaller pods of 2 or 3 are more common. The southern journey, which extends well into the winter months, brings the whales close to land outcroppings all along the Pacific coast, and it is during this portion of the migration that whale watching hits its peak.

When swimming on the surface during migration, a gray whale normally takes a series of 3 to 5 breaths about twenty seconds apart before dipping its head and throwing its tail into the air to begin a 3 to 5 minute dive that can easily reach a depth of several hundred feet.

In early winter, the whales reach the waters off the Pacific coast of Baja. Most of the animals stop at Scammons Lagoon, San Ignacio Lagoon, or Magdelena Bay, a series of shallow bodies of water along the west coast of Baja, but some gray whales go all the way to the Sea of Cortez. Scammons Lagoon, San Ignacio Lagoon, and Magdalena (often called "Mag") Bay are large, well protected areas tucked into the

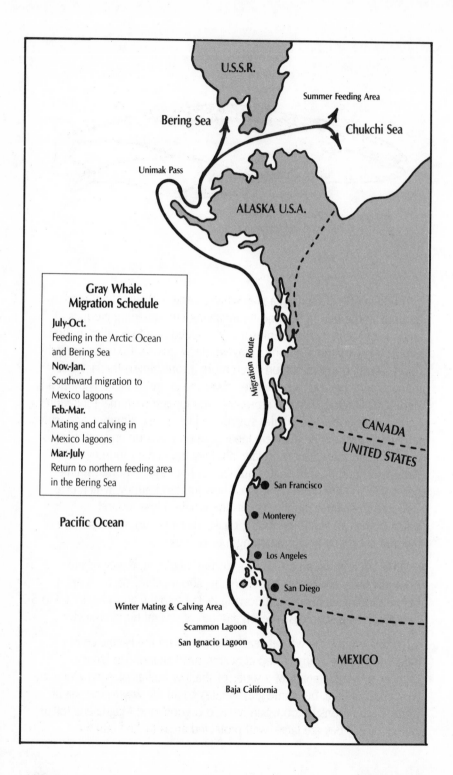

U.S.S.R.

Summer Feeding Area

Bering Sea

Chukchi Sea

Unimak Pass

ALASKA U.S.A.

**Gray Whale
Migration Schedule**

July-Oct.
Feeding in the Arctic Ocean
and Bering Sea
Nov.-Jan.
Southward migration to
Mexico lagoons
Feb.-Mar.
Mating and calving in
Mexico lagoons
Mar.-July
Return to northern feeding area
in the Bering Sea

Migration Route

CANADA

UNITED STATES

San Francisco

Pacific Ocean

Monterey

Los Angeles

San Diego

Winter Mating & Calving Area

Scammon Lagoon

San Ignacio Lagoon

MEXICO

Baja California

188

seclusion of the Baja wilderness. It is there that pregnant females give birth and where many courting whales successfully mate.

Mating, which produces a single calf, usually occurs in groups of three, one female and two males. The role of the second male is a subject of some debate. Some authorities maintain that the second male is present only to learn the finer points of courtship, while others suggest his purpose is to steady the female so the first male can achieve penetration. The courting ritual takes place on the surface, and often lasts for several hours.

The calves are born in early winter after a 13 month gestation period. At birth, calves are between 12 and 17 feet long, and weigh up to 3,500 pounds. While that certainly seems large for starters, it is equally true that the calves are capable of gaining up to 220 pounds per day for several months following birth. Cows are highly protective of their offspring and keep their young close to their side for 2 or 3 months after birth.

An obvious question to ask, is "why do whales travel 5,000 miles to calve and breed." The answer, though not completely understood, centers around the fact that the lagoons offer a habitat in which the calves have the best chance for survival. Not too many years ago, grays utilized more northern regions such as San Diego Bay, but heavy shipping traffic has forced them to seek quieter waters. How far they can be pushed without greatly endangering the species is a point of great concern.

In late winter and early spring, the whales begin their northern journey toward the feeding grounds. This route takes them further out to sea than does the southbound segment. Exactly how grays successfully navigate such great distances is a point of scientific contention. It is not known whether they rely primarily upon sight, hearing, echolocation, or a combination of all three. Gray whales are often seen "standing on their tails," or more precisely holding their heads out of water in an act called spy-hopping, sparring, or spying out. Some authorities believe the whales are trying to visually orient themselves with prominent landmarks, while others believe the grays' out-of-water vision to be rather poor.

Throughout the migration, gray whales can be seen "leaping" almost entirely out of the water in a manuever called breaching. Swimming at speeds of close to 30 miles an hour before propelling themselves skyward, the whales often spin around in the air before crashing and splashing their way back to the surface. This spectacular action is often

Baleen Whales

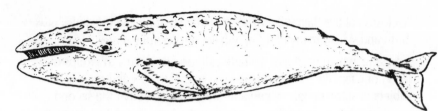

California Gray Whale

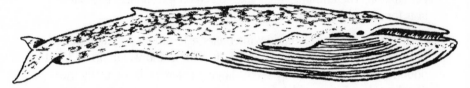

Blue Whale

Finback Whale

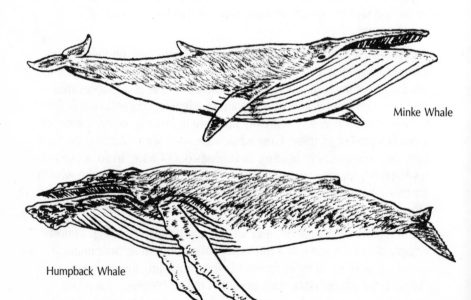

Minke Whale

Humpback Whale

repeated as many as 10 times in succession within a 2 or 3 minute span. The purpose of breaching is not known, as some scientists believe it is for fun, others to dislodge parasites, and still others maintain that breaching is a response to confusion or the feeling of being threatened. A recently proposed theory states that breaching serves as a rather elaborate part of communication between whales.

Underwater encounters with these clever creatures are rare, especially during migration. In spite of their size, approaching gray whales will almost always turn or dive to avoid a confrontation with divers. To my knowledge, I am one of a small number of divers who has been fortunate enough to repeatedly swim eyeball to eyeball with wild gray whales. I have done so only twice in California waters, but have spent considerable time with grays in the waters of Baja while working on documentary film projects.

During those dives, I learned that it was not so much their size as their incredible body control and grace which impressed me the most. Grays can repeatedly swim within inches of a diver without making contact. However, when threatened they can, and will, accurately and powerfully throw their flukes at an intended target. Considering their size . . . well, enough said.

Their size, however, is overwhelming. Underwater, you soon realize that grays have body parts that are bigger than you are. Their tails alone can be 10 feet across. To see a gray approach you while you are bobbing on the surface is to watch your horizon change shape as the top of the whale's head blocks out a portion of the sky. That sight is a diving thrill of a lifetime.

While most people are not likely to ever spend a minute with a gray whale while underwater, just being able to observe them on the surface greatly enhances our appreciation of the marine world.

Blue Whales

Blue whales (*Balaenoptera physalus*) are the largest of all whales, and are in fact, the largest living animal on earth. They are reported to 111 feet in length and weigh in excess of 100 tons. At birth blue whales are over 20 feet long, and weigh almost 3 tons. By the time they are 8 months old, they can attain a length of 50 feet. Like most of the oceans' largest creatures (blue whales, finback whales, humpback whales, whale sharks, and basking sharks), blue whales are filter feeders, preying primarily on krill and other planktonic life forms. The folded ridges on the underside of their head allow blue whales to open their mouths extremely wide as they swim through dense concentrations of food. Blue whales can be positively identified by the combination of their size, and the fact that they are the only deep blue to jet black whales that have the folded ridges on the underside of the jaw.

Next to gray whales, blue whales are the second most commonly observed baleen whale in California waters. Once hunted almost to the point of extinction, blue whales have made a surprising comeback. It is generally believed that the blue whale populations of the northern and southern hemispheres never mix.

Finback Whales

Finbacks (*Balaenoptera physalus*), or fin whales as they are sometimes called, are probably the most commonly seen whale off the California coast during the summer. Similar in build to blue whales, fins reach lengths of 76 feet. Thought to be the fastest swimmer among the larger whales, finbacks are easily recognized from the surface by the small dorsal fin which is located nearer the tail than one might normally expect. Like most of the larger whales, finbacks are baleen whales.

Minke Whales

Minke whales (*Balaenoptera acutorostrata*) look a great deal like miniature finback whales. The largest minke ever documented was 33 feet long. At birth minkes are generally between 7 and 9 feet in length. Rather slow swimmers, minkes are thought to be solitary whales, especially when they swim close to shore. Minkes feed on small crustaceans, and some small fishes.

Humpback Whales

Humpbacks (*Megaptera novaengliae*) are easily distinguished by the comparatively long pectoral flippers which are knobbed along the forward edge. The flippers can be almost 15 feet long, while the entire animal reaches a length of just over 50 feet. The flippers stand out prominently during dramatic breaches, as the humpbacks leap clear of the water, spinning in a circle, before they crash back to the surface.

The "songs" of the humpbacks have attracted considerable attention in recent years. Their pattern of vocalization is easily distinguished from the sounds of other species of whales, and is believed to help humpbacks communicate for distances of several hundred miles.

California Marine Mammals

Like many pinnipeds, elephant seals are highly ꝏcal.

2. Sea otter.

8. California sea lion pup shows off its form for the author's camera.

4. California sea lion pup.

5. A male California sea lion keeps close watch of his females and newborn pups.

6. A female elephant seal barks out an emphatic "no" to the amorous advances of a bull.

Social interaction is an integral part of the lives of elephant seal pups.

. Harbor seals appear curious, but they tend to be wary around humans.

9. Herds of bottlenose dolphins often "surf" the bow wake of boats in California waters.

10. A California gray whale calf gets a boost toward the surface from its mother.

1. Common dolphins feed on a wide variety of fish and invertebrates in California waters.

2. Male killer whales can be easily identified from the surface by their highly conspicuous dorsal fin which can tower as high as 5 feet above their back.

California Marine Birds

13. Western gull.

14. Once threatened by extinction, brown pelican are making a remarkable comeback.

15. Heermann's gulls can be distinguished from other gulls in California by their darker coloration and brigh red beaks.

6. Marbled godwits.

7. Snowy egrets inhabit many California marshes.

18. Great blue heron.

9. Herring gull aloft near Catalina Island.

20. Pelagic cormorant.

21. Elegant tern.

22. California's marine birds play an integral role in the marine environment.

Toothed Whales

Pilot Whales

Pilot whales (*Globicephala macrorhyncus*), sometimes called blackfish, are fascinating animals. They usually travel in pods of up to 50 whales. Reaching documented lengths of 22 feet, the dark brown to black coloration and bulbous, melon-shaped head make them easily recognizable. Their favorite food is believed to be squid. During large squid runs at the Channel Islands and in coastal canyons, pilot whales are quite commonly observed in large pods as they feed upon squid.

Common Dolphin

The white, hourglass shaped belly of common dolphin (*Delphinus delphi*) allows for easy, positive identification. Traveling in large pods during late summer, fall, and early winter, common dolphins are often seen leaping entirely out of the water as they travel. Fish feeders, these dolphins are relatively small, attaining a maximum length of 7 or 8 feet, but they are exceptional swimmers. In the wild, common dolphins have been recorded swimming up to 18 miles per hour.

Common dolphins range from British Columbia to Ecuador, and like many other toothed cetaceans they feed primarily upon small fishes.

Bottlenose Dolphin

As the name suggests, bottlenose dolphins (*Tursiops truncatus*) have a prominent, bottle-shaped beak. Attaining a length of up to 12 feet, these dolphins are generally larger than common dolphins. Bottlenose are an almost uniform gray color, and are stout bodied, weighing as much as 800 pounds. Feeding upon a variety of fish, bottlenose dolphins are known to consume up to 20 pounds of mackerel per day while in captivity. This species is one of the best known and most widely written about of all dolphins. The subject of extensive research, bottlenose dolphins possess large, convoluted brains. Some scientists believe that next to man, dolphins are the world's most intelligent animals. Having

Echolocation in Marine Mammals

In most mammals vision is rather well developed. Even most marine mammals are believed to have a very useful sense of vision underwater — even though their ability to see through air is suspect. In addition, several groups of mammals including some whales, dolphins, and pinnipeds have developed the ability to orient themselves to their surroundings by producing sharp sounds and listening for the reflected echoes. This phenomenon is known as echolocation. Although it is difficult to be absolutely sure of all the mammal species that use echolocation, approximately 20% of marine mammals are believed to possess the ability to echolocate. Those species that do are quite capable of using their ability to echolocate to find objects, even very small objects, in total darkness.

The sounds that are utilized in echolocation should not be confused with those used in "songs" of communication, such as the well known song of humpback whales. The sounds most useful for echolocation are trains or pulses of clicks of very short duration. As an example, bottle-nosed porpoises use a series of clicks, with each click lasting only a portion of a millisecond. The clicks are often repeated up to 800 times per second. The porpoise then waits for the sound wave, or echo, to bounce off of an object and to return before emitting another series of clicks. The porpoise can gauge the distance to objects and measure the speed at which they are moving by measuring the time required for the echo to be returned. In addition, in tests conducted with porpoises in captivity, the animals have repeatedly shown an incredible ability to distinguish between objects of similar shape and size, but that have a different density.

Some of the clicks are well within the range of human hearing, while other clicks are of a higher frequency than humans can discern. It is generally believed that low frequency clicks serve to orient in a general sense, and that higher frequency clicks are used when finer discrimination is required.

Toothed Whales

Pilot Whale

Common Dolphin

Bottlenose Dolphin

Orca

spent some time in the wild with bottlenose dolphins and some of their close cousins, I have to wonder if those rankings would be fair to the dolphins. They are truly remarkable animals. Like many dolphins, bottlenose dolphins use complex patterns of whistling as well as body language in order to communicate. Dolphins also utilize sonar to navigate, explore their surroundings, and to help in the search for food.

Orcas or Killer Whales

Orcas (*Orcinus orca*), or killer whales as they are often called, are the largest members of the dolphin family. Without question, they are the most widely misunderstood, and certainly the most misrepresented of all whales. It is true that killer whales are the only cetacean that is known to commonly prey upon other warm blooded animals, but attacks on humans remain virtually unrecorded. It is well documented that orcas feed upon seals, walruses, birds, dolphins, and other whales, including blue whales. They are also known to prey upon squid and fish. While hunting, members of the pod often cooperate in a communal hunting effort, a behavior that has been witnessed on numerous occasions as orcas hunted both blue whales and sperm whales.

The commonly used name, killer whale, is derived from an errant translation of an Eskimo term. Years ago, the Eskimos referred to men that hunted orcas as "whale killers." The English translation mistakenly became "killer whales," a term that within the confines of English described the whales, not the hunters.

Killer whales are distinctly marked, having a black body with white underparts, and a large white spot just behind and above the eye. Mature males possess a large, triangular dorsal fin which stands as high as 6 feet off the back of the whale, making them easy to identify when seen from the surface. Males have been measured to be more than 30 feet long, and are unusual among cetaceans in that mature males are much larger than females.

Killer whales are large, powerful predators that are capable of rapid bursts of speed, but having been in the open ocean with a killer whale, I am convinced, as are many others who have had similar experiences, they are by no means indiscriminate killers. In fact, they generally prove to be somewhat shy and difficult to approach. In some instances the whales have demonstrated some curiousity about divers, but aggressive behavior in the wild toward divers is almost unheard of.

196

Pinnipeds

California waters are commonly inhabited by seven species of seals and sea lions, a group of animals collectively referred to as pinnipeds by the scientific community. The order Pinnipedia (fin feet) is subdivided into 3 families; (1) Otaridae which includes sea lions and fur seals, (2) Phocidea which consists of the true seals (hair seals), and (3) Odobenidae, the walruses, which are found only in Arctic waters.

Sea lions and seals share many common characteristics. All of these animals have thick hides with heavy layers of fat underneath which protects them against their cold surroundings. Both groups have modified fore and hind limbs called flippers, that help them manuever both in and out of water.

However, significant biological differences do exist between sea lions and seals. In fact, in strict scientific terms, some experts maintain that seals are more closely related to both cats and bears than they are to sea lions. This distinction is made because it is believed that in the process of evolution seals, cats, and bears split from a common ancestor, while sea lions did not evolve until the "tree of ancestors" had been developed further.

In the field sea lions can be distinguished from seals by noting three factors. First, sea lions possess small external ears, which seals lack. However, to clarify a point of potential confusion, it is worthwhile to note that sea lions are often correctly referred to as eared seals. Second, the two groups can be separated by observing differences in the size and use of their flippers. The foreflippers of sea lions are comparatively large, and are used as the chief means of propulsion in the water. Sea lions use their hind flippers as a rudder, while seals propel themselves with their rear flippers. Seals are unable to turn their hind flippers forward, and are, therefore, less mobile on land than are sea lions. And third, sea lions have a harsh coat, while fur seals possess a dense, soft undercoat that is protected by coarse, guard hairs.

Both seals and sea lions are considered to be highly social animals, gathering in large herds or colonies at many times during the year. Pups of both sexes and females without pups are more gregarious than breeding males and females who have recently pupped.

It is quite common to see different species of seals and sea lions mixed in with one another both in the water and on land. In fact, during the spring at Point Bennett on San Miguel Island, the area is inhabited by upwards of 10,000 specimens consisting of 6 different species of pinnipeds. That gathering represents the largest number of marine mammal species gathered at one time anywhere in the world.

California Sea Lion

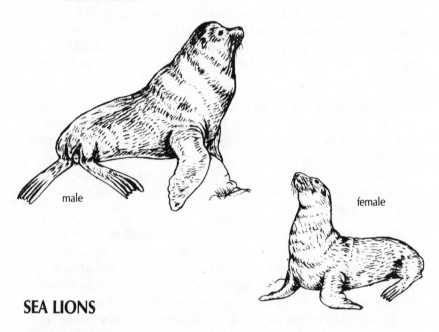

male   female

SEA LIONS

California Sea Lions

By far and away, the species of sea lion that is most commonly seen in California is the California sea lion, a species which is scientifically named *Zalophus californianus*. The bulls of the species weigh as much as 700 pounds and attain lengths of better than 10 feet. Males are generally much larger than females and have a noticeably larger head due to the pronounced sagittal crest, an arrow shaped ridge extending from the forehead to the rear of the skull. Mature females are more lightly colored than males, weigh close to 250 pounds, and reach lengths of close to 8 feet.

Throughout the year, whether during their breeding season or not, California sea lions are quite gregarious, preferring to gather in groups. This behavior is especially true of females and pups, evidenced by the fact that during the non-breeding season of August through April, herds of sea lions huddle closely together on land when they haul out of the water. During these months their social structure is, however, much more loosely organized than in mating season.

When seen on land, sea lions are most often found in one of two areas; (1) a rookery, which is defined as a coastal region occupied by breeding populations, or (2) a hauling out site, a strip of land occupied by non-breeding animals. On land, sea lions appear awkward, but on the rocky and slippery terrain of the California coast, bulls can often move faster than humans for short distances. Bulls are highly protective of the territory, and just by keeping their size and your size in mind, nothing more need be said about approaching them too closely.

During the mating season (approximately May through July), the social behavior of California sea lions changes considerably. Sea lions are polygamous, meaning that both sexual partners have more than one mate. Breeding males become strongly territorial and vigorously defend their realms. Males claiming a territory bark incessantly in a communication intended to inform competing males not to enter their domain. Confrontations between competing males can be rather intense, with the fights involving a lot of chest-to-chest pushing and biting. The goal is to physically push the other male out of the territory. Extensive scarring is a common sight on the chests of adult males.

While the ratio of females to bulls in any one area might be as high as 14 to 1, it is interesting to note that bulls compete only for physical territory, and make no effort to prevent females from breeding with bulls in other areas. Copulation occurs both on land and in the water. The female will bear the young approximately 12 months later during the next breeding season, although the gestation period is considered to be only nine months. The difference in the two time periods is due to a phenomenon called delayed implantation, where the fertilized egg lies dormant in the uterine wall for three months, enabling females to give birth at the rookeries where they breed.

When entering a rookery to bear young, pregnant females tend to band together. Females can be quite aggressive in the defense of both themselves and their young just prior to and after giving birth. Mothers

Snorkeling and Diving with Sea Lions

On perfect diving days all along the California coast, when the weather, water, and sea life cooperate, divers simply cannot have more fun than when diving near a sea lion rookery. During those magic days, the sea lions appear as fascinated by you as you are with them. They almost never leave your side. An entire group might pause upside down directly in front of you, almost within touching distance, as they stare into your mask. Perhaps they will swim circles around you, or blow a faceful of bubbles at you as they dart in and out providing entertainment rarely equalled in the marine kingdom.

On rare occasions, sea lions will even allow divers to reach out and touch them during a game of underwater chase. But make no mistake about it, in such a contest, the sea lions are clearly in control. You are "it", the tagger until they allow themselves to be tagged. Without question, their superior grace, speed, and coordination make them the envy of their human counterparts. As divers, we are truly fortunate to be able to watch them perform their underwater choreography in California.

At other times, I have been absolutely certain that I was surrounded by sea lions, yet, underwater I never even caught the slightest glimpse of one. I saw them on the beach, watched them scramble into the water as I approached, and then vanish without a trace. During times like these, pursuit is almost laughable. You will quickly learn that an encounter between divers and sea lions is totally up to the discretion of the sea lions. When they choose to avoid you, the best you can do is hope tomorrow will be different.

In the water, the design and movement of sea lions is most astonishing. There, sea lions display a form and proficiency that is truly spectacular. Supremely skilled as swimmers and body surfers, sea lions are perfectly adapted for life along the coast. Whether pursuing prey, turning sharply to avoid being touched, or surfing in waves that are much too large for humans to enter, sea lions always appear to be in total control.

Pups are usually the most curious about divers. They are often enthusiastic in their play, occasionally nipping at divers and their equipment. It is all in good fun, of course, but it can be somewhat unnerving the first time it happens to you. Females and non-breeding males also display a sense of playfulness and curiousity, but bulls tend to keep their distance, approaching only close enough to check out the situation unless they feel their territory has been trespassed upon. When threatened or encroached upon, bulls have been known to rush toward divers and blow bubbles in a threat display, though overt aggression causing injury to divers is not common.

However, during any time of year when you are diving with a game bag full of abalone or fish, the entire game changes. Sea lions like to eat too, and have been known to have rather poor manners when it comes to stealing a diver's catch.

In natural settings, I have no doubt that sea lions can stand up for themselves when the need arises. On numerous occasions while out filming sharks in the deep waters off the Southern California coast, I have watched sea lions compete with both blue sharks and mako sharks for bait we had placed in the water. Though I was quite surprised initially, I have repeatedly observed sea lions biting and chasing away both blues and makos, while at the same time almost always ending up with the bait. It is interesting to note, that while great white sharks are known to feed on seals, it is extremely rare (only 3 instances by 1985) to find a documented case of a great white shark taking a sea lion. Sea lions have much greater manueverability than both sharks and seals.

and pups are quick to develop a strong relationship and have a keen ability to identify one another very early in the pup's life. During the first few weeks after giving birth, mothers rarely leave the pup's side, even when entering the water to escape the heat.

Mothers help their offspring learn to swim by a modified sink-or-swim method. The mother must usually drag a somewhat less than enthusiastic pup into the water, and then support and push the pup through the awkward stages. Although the first few minutes appear to be a little frantic for the pups, they quickly take to water.

When on land, sea lions are usually found close to the water's edge, often within splashing distance of the waves. Sea lions are extremely sensitive to heat, as the layer of subcutaneous fat can quickly cause internal body heat to elevate to dangerous levels. For that reason, sea lions rarely stray far from the water.

On land, sea lions can be approached easily if you stay downwind, moving slowly while keeping a low profile. This fact does not mean that sea lions are comfortable with humans, but more likely indicates that their senses on land do not allow for ready detection of predators. If startled by sudden movement or by approach from above, entire herds sometimes stampede. During such a frantic time, sea lions tend to dash rapidly for the water and, on occasions, many suffer serious injuries or fatal falls in their haste.

Sea lions typically feed on squid and small fish. Though they are considered to prefer nocturnal feeding, it is not uncommon to see them feed during daylight if the opportunity arises. In fact, their feeding habits have made all sea lion species the center of controversy in California. Sea lions like to feed upon many species which are also pursued by both commercial and sport fishermen, and it is well known that sea lions will readily steal a fisherman's catch if possible. Many, though certainly not all, fisherman retaliate by shooting at or bombing the sea lions. Doing so is strictly a short term solution. While it is obviously frustrating to lose a catch, in the long run it is also known that sea lions feed upon other animals that prey upon fish populations. For example, in northern waters sea lions feed heavily on lamprey, which in turn inflict extensive damage on salmon populations, especially when their numbers are not naturally controlled.

Stellar Sea Lions

Stellar sea lions (*Eumetopias jubatus*) inhabit California waters, though they are not nearly as numerous as California sea lions. The largest colonies are in Northern California at the Farallon Islands and at Ano Nuevo. In the early 1980's the total population was estimated at close to 1,700 animals. Stellar sea lions attain much greater sizes than do California sea lions. Reaching lengths of up to 13 feet, bulls weigh as much as 2,200 pounds. Females reach 9 feet, and weigh just over 600 pounds. Stellar sea lions are a yellowish-brown in contrast to the darker brown of California sea lions. California sea lion bulls possess a prominent saggital crest which is lacking in Stellar males. When on land, other than when breeding, Stellars are typically much quieter than California sea lions who seem to bark endlessly. Like California sea lions, Stellars feed primarily on fish species that have little or no commercial value and on squid.

Stellar Sea Lion

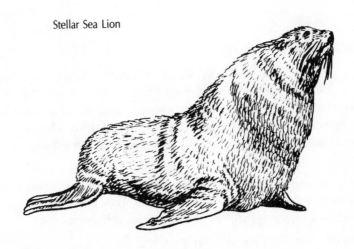

SEALS

Four prominent species of seals are known to commonly inhabit California waters. They are harbor seals, northern elephant seals, Guadalupe fur seals, and northern fur seals, sometimes called Alaskan fur seals. Of these, harbor seals are the ones most likely to be seen by divers, but all can be seen by boaters and fishermen.

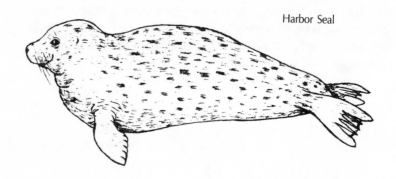

Harbor Seal

Harbor Seals

In addition to the previously discussed features which can help you distinguish sea lions from seals, harbor seals (*Phoca vitulina*) are easily identified by their small size, chunky shape, small front flippers, and dark spotted coats which have a light colored base. Males reach sizes up to 6 feet in length, and weigh up to 300 pounds, while the females are slightly smaller.

Harbor seals are considered to be polygamous or promiscuous, and do not form organized harems when breeding. They prefer to feed upon squid, octopi, some shellfish, and numerous fish species.

In the water harbor seals (*Phoca vitulina*) appear more wary of intruders than do sea lions, rarely coming much closer than 15 or 20 feet.

Harbor seals range from Alaska to San Geronimo Island, off the coast of Baja.

Northern Elephant Seal

Northern Elephant Seals

Although more properly referred to as northern elephant seals, Californians typically omit the word northern, and refer to the species simply as elephant seals. The male elephant seal (*Mirounga angustirostris*) is the largest of all the seals, reaching a length of 16 feet and weighing up to 5,000 pounds. Males develop a large, bulbous snout. Elephant seals derive their name from the combination of (1) their massive size, and (2) the males' snout. Females lack the large snout, but like the males, are brownish to silver gray. The females attain a size of up to 11 feet long, and at that size the females weigh just under a ton.

Elephant seal populations were once quite abundant along the California coast, but as with so many species of pinnipeds, they were hunted to the point of near extinction. They have, however, made a very encouraging comeback both in Baja and at the California Channel Islands, where breeding colonies can be found at San Miguel, San Nicolas, and Santa Barbara. Breeding colonies also are present at the Farallon Islands and at Ano Nuevo. By 1980 the California count had exceeded 50,000 elephant seals.

Nocturnal feeders, elephant seals prey primarily in deep waters upon rays, ratfish, rockfish, squid, and small sharks. Great white sharks are believed to prey heavily upon elephant seals when the opportunity presents itself. Elephant seals are only rarely seen by divers.

Northern Fur Seals

One of the best known of the seal species on a worldwide basis, northern fur seals (*Callorhinus ursinus*) have been valued for years for their furs. Preferring to congregate in large numbers at their rookeries, these animals have been hunted to the point of near extermination. The original worldwide herd was estimated to have approximately 2,000,000 specimens, but before the international treaty of 1911 which brought these seals significant protection, their numbers had diminished to only 125,000. By international agreement, northern fur seals are still hunted. But you'll be glad to know the species has made a remarkable comeback, and by 1979 the population was estimated to exceed 1.5 million.

When breeding season nears, the bulls are the first to head for the rookeries. Bulls fight fiercely for their territory, which might be as small as 40 square feet. Harems usually consist of around 40 females. The pregnant females arrive last. Soon after, they give birth, and only one week later they are willing and able to breed again.

Like most pinnipeds, northern fur seals almost always bear only one pup. After giving birth, the mother feeds the pup heavily, and then leaves to go to sea for intense feeding for a week or longer. Upon returning, the mother nurses her pup for 2 to 3 months before leaving it on its own. Breeding males remain on land during that time without eating or drinking, living solely off of stored fat.

Northern Fur Seal

Males are dark brown, with a gray neck and shoulders, while females tend toward the gray side. Males are considerably larger than females, reaching sizes of 8 feet and 700 pounds. Females peak out at about 5 feet and 130 pounds. These seals reach sexual maturity in 3 years.

Northern fur seals prey upon a wide variety of food sources, including anchovies, squid, hake, and herring. Other than man, they are sought after by sharks and orcas. In addition, parasitic worms often inflict great damage on pups.

Guadalupe Fur Seal

Guadalupe Fur Seal

Like so many pinnipeds that reside in California, Guadalupe fur seals (*Arctocephalus townsendi*) have been hunted almost to extinction. Once they ranged from San Benitos Island off the coast of Baja to the Farallons in Northern California, but they were so heavily pressured that for some time they were thought to be extinct. While Guadalupe fur seals have been sighted in recent years at the Channel Islands, as a species they continue to fight the odds for survival.

Not as much is known about Guadalupe fur seals as is known about many pinnipeds, though they are believed to have a great deal in common with northern fur seals. Both the males and females are dark brown to blackish-gray, with lighter gray heads and necks. The long pointed muzzle, and the silver mane created by long guard hairs bordering the neck and shoulders provide reference points for positive identification. The male attains a length of up to 6 feet, while the females are slightly smaller.

Weasel Family

Sea Otter

Sea Otters

Members of the weasel family, sea otters (*Enhydra luris*) are unquestionably some of the most interesting of California's marine mammals. In an evolutionary sense, sea otters are among the youngest of the marine mammals, being more closely related to their terrestrial cousins than most marine mammals. Though otter populations are much more heavily concentrated in the northern part of the state, these mammals are occasionally seen all the way to the Mexican border.

Sea otters are noted for their reddish brown to black fur, which is dense, soft, and very fine. Many older males are white headed, but specimens of both sexes and varying ages also display this quality. Males reach lengths of close to 4½ feet, including the 12″ long tail, and can weigh close to 85 pounds. The females are considerably smaller. They reach lengths of up to 4 feet and weigh only 60 pounds.

Otters are well known for their ability to utilize their stubby, rounded forepaws to grasp rocks. In turn, the rocks are employed to puncture holes in abalone in order to free them from the rocky substrate. Otters also use rocks by positioning them on their chests so as to grasp hard shelled food sources such as crabs, and bash them against the rocks to make the meat accessible. In addition, otters are known to feed upon snails, mussels, squid, octopus, tubeworms, limpets, barnacles, starfish, chitons, and clams. Sea otters rarely prey upon fishes.

Sea otters are most often seen on the surface in kelp beds either resting on their backs or swimming at a leisurely pace. However, they are quite capable as swimmers, and are known to make dives last more than 4 minutes. The hind feet of sea otters are webbed, being used for thrust when swimming. The tail, too, is used as a swimming aid.

For thousands of years otters naturally inhabited waters between Alaska and the lower portion of the Baja Peninsula. Their most viable natural predator is believed to have been sharks, but studies show that otter populations were quite prolific. Then, as has so often been the case, along came mankind. Hunters and trappers began to pursue otters in quest of their prized pelts. In the late 1700's, hunters from several European nations, Russia, and America began to exploit the economic value of the pelts. Throughout the 1800's and early 1900's the pressure continued as the sea otters proved to be an easy prey. It is estimated that in 170 years of hunting, over 1 million otters were killed, and sadly the otter populations declined to the point of near extinction. Towards the end of that era a series of international treaties were agreed to, and by 1913 sea otters were finally protected in international waters, as well as within the Alaskan territory and the region governed by the state of California. It is interesting to note that native Indian populations all along the west coast had successfully co-existed with otters for centuries, even though they hunted them for both food and furs.

Presently the sea otter population in California is concentrated in a 200 mile stretch between Santa Cruz County to the north to San Luis Obispo County to the south. While most individuals remain within a home territory, the range has been expanding at about 2.5 miles annually, and the total population within California has been increasing by approximately 5% per year.

Sea otters can be regularly observed in a number of locations in Central California, especially along the Monterey Peninsula. If there is a lot of bull kelp in the area, spotting the otters is more difficult than when giant kelp is predominant. The dark brown floats of bull kelp bear strong resemblance to otter's heads and bodies. The best way to locate the otters is to watch carefully as the swells roll through. When the swells reach a high point, the bull kelp is often submerged, but the otters remain on the surface. Binoculars will prove valuable. The presence of gulls and other sea birds that hover over feeding otters can be used as an aid as well.

Underwater, otters provide superb diving entertainment. Some days they prefer to keep a distance, but there are days when their natural curiosity prevails. Feeding upon animals that are typically found in relatively shallow waters, otters do not usually dive much deeper than 100 feet, but they have been documented to reach depths in excess of 300 feet in efforts to secure food.

The Sea Otter Controversy

Today, sea otters play a vital, yet quite controversial role in the ecological balance of the state's water. They are voracious eaters, feeding on a wide variety of sources, including abalone, lobster, sea urchins, crabs, snails, mussels, squid, octopi, scallops, and more, though they rarely eat fish. Lacking a fat layer to assist in maintaining their 100° F body temperature, young otters often consume up to 35% of their body weight in a day's feeding in order to combat the effects of cold water. Adults eat closer to 15% of their weight every day. Translated into annual terms, that means an adult otter devours close to 5,000 pounds of food a year, and the entire population will consume in excess of 6,000 tons. Furthermore, wild otters are documented to live to be at least 20 years old. It is easy to conclude that as a result, large otter populations limit the size of abalone, lobster, and other shellfish populations through their heavy feeding. Obviously, that fact presents a considerable problem for commercial fishermen, many of whom lobby strongly against the reintroduction of otter populations.

However, otters also feed heavily upon sea urchins. Without the otters, sea urchin populations increase dramatically. As has been mentioned in several places in this book, sea urchins play an integral role in the overall health of kelp forest and reef communities. Urchins normally prefer to feed upon kelp shed. However, when urchin populations become extremely dense and competition for food intensifies, urchins will readily feed upon the holdfasts of living kelp plants. The urchins eat through the holdfasts, weakening them to the point that the holdfasts pull loose. That kelp plant dies, and in a "domino effect" often becomes entangled with other plants, pulling them loose from the bottom. As a result, these plants also perish and they often become entangled with still more plants as the cycle continues. Without the kelp, the mainstay of the kelp forest ecosystem, the entire food chain suffers extensively.

As a result of the combination of their protected status, varied diet, foraging abilities, and the general state of well being of California waters, sea otter populations have made a substantial comeback. However, because of new threats and the fact that the population has not recovered to its original size, sea otters continue to be classified as an endangered species.

California
Marine Birds

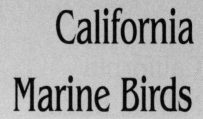

California
Marine Birds

The existence of many species of birds greatly affects the ecology of California's marine environment. Birds play a major role in the link between land and sea. Many species of birds prey upon a variety of sea creatures including fishes, crabs, worms, mussels, shrimp, and squid. In turn, many land animals prey upon birds. Some species of birds also play a major role in the propagation and distribution of fishes by unintentionally transporting eggs from one locale to the next.

In turn, many birds are dependent upon coastal marshlands, bays, estuaries, tidal flats, and harbors for nesting grounds. Unfortunately, as the influence of the industrialized world spreads, many acres of breeding ground are being irreversibly destroyed. That situation presents a major problem to the survival of numerous birds, as well as the many plants and animals with which the birds interrelate.

Over the years ornithologists have attempted to classify birds in a variety of ways. One method of categorizing birds is by habitat preference. California's seacoast extends for a length of nearly 1,100 miles, and obviously includes a diversity of habitats. As far as marine birds are concerned the major subdivisions, by habitat, are as follows:
1) the open sea
2) the offshore islands, and coastal seacliffs
3) the beaches, reefs, and coastal waters
4) the lagoons, bays, and estuaries, and
5) the tidal flats and salt-water marshes.

The open sea is generally defined as water that is out of sight of the mainland. True pelagic birds do not come ashore except to roost, meaning many species of pelagic birds spend up to 10 months a year at sea. These species include shearwaters, storm-petrels, jaegers, many alcids, and albatrosses. The birds of the offshore islands and coastal seacliffs include brown pelicans, three species of cormorants, western gulls, and black oystercatchers. In addition, grebes, loons, scooters, and phalaropes are coastal birds, but they rarely, if ever, come ashore. Some species such as storm-petrels nest both at the islands as well as on the mainland.

The coastal beaches represent a dividing line between coastal waters on one side and tidal estuaries and marshes on the other. The beaches are frequented by many species at one time or another and are commonly visited by gulls, godwits, sanderlings, willets, long-billed curlews, and some species of plovers. The shorebirds of the lagoons, bays, tidal flats, and marshes include herons, egrets, many species of gulls, ducks, geese, godwits, and curlews.

In general conversation, people tend to refer to those birds that are commonly seen in the just discussed biomes at one time or another during the year as marine birds, or California marine birds. However, doing so is often a matter of convenience. Writing a strict definition of a marine bird is a confusing task. Many of the species that are often seen near California waters are also seen as far inland as Iowa, as far east as Europe, as far north as the Arctic Circle, and as far south as the southern tip of South America. Obviously, labeling those species as "California marine birds" can be rather misleading. Some of these "marine birds" spend 10 months a year on the water, others spend most of their lives over land. Some species sleep on the water, and others rarely touch it, except when seeking food. Some can safely drink salt water, while others can not. In the final analysis, it is rather easy to understand why it is such a difficult task to write an accurate decription of a marine bird.

Regardless of strict definitions, many of us enjoy watching the variety of birds that can be observed along the California coast and out to sea. Without having to become a credentialed ornithologist, it is fun to be able to identify at least some of the more common species and to enjoy a basic knowledge of their natural history. This section of the text is included in order to make you more familiar with some of the more prominent birds of the California coast.

Realizing that the various habitats exist in the language of birders should help you utilize other reference sources when seeking additional information. However, because birds cross these boundaries so often, and in many cases for such extended periods, I have chosen to group the birds according to their families.

There are at least 37 families of marine birds represented along the California coast and over oceanic waters including representatives from the following families:

Albatrosses	Loons
Alcids	Mergansers
Avocets, Stilts	Oystercatchers
Boobies and Gannets	Pelicans
Calidridine Sandpipers and allies	Phalaropes
Cormorants	Plovers
Cranes	Rails, Gallinules and Coots
Dabbling Ducks	Shearwaters, Petrels
Dippers	Skimmers
Diving Ducks	Snipe, Dowitchers
Frigate Birds	Still-tailed Ducks
Geese	Storks
Grebes	Storm Petrels
Gulls	Swans
Herons, Egrets and Bitterns	Terns
Ibises, Spoonbills	Tree Ducks
Jaegers, Skua	Tringine Sandpipers and allies
Kingfishers	Tropicbirds
	Turnstones

The following information provides physical descriptions of adults and brief natural histories of some of the more prominent families and species. Like many other types of animals, many species of birds undergo several different appearances during the various stages of their lives and during different times of year. This guide deals primarily with a basic description of the adults.

It is worthwhile to note that unlike the common name of many other species of animals, the common name of a bird is recognized as the correct or "official" name by scientists and ornithologists.

214

Identifying Birds in the Field

Positively identifying birds in the field can be quite difficult to do when you first try your hand. Like many other endeavors that vary from riding a bicycle to learning to type, once you practice a little it becomes difficult to remember exactly why the task seemed so overwhelming in the first place. Bird watching is similar to bike riding and typing in that respect.

If you are new to bird watching and you see a bird you want to identify, the following suggestions should help you get started by serving as keys to identification:

(1) Look at the overall shape and size of the bird in question. Mature eagles and sparrows obviously have different builds, and even inexperienced bird watchers have no trouble in discriminating between their distinguishing characteristics.

(2) Look for feather patterns. Overall color can be useful, but remember plumage often varies throughout the year, from mature to immature bird, and by sex. Patterns of dark and light feathers can, however, be very helpful as an identification key.

When using color, instead of looking for one overall color, try checking for the color of the breast, feet, rump, and bill. The presence or absence of a crown stripe on the head is a good identification key. The color of the crown stripe, and the presence and color of an eye stripe should be noticed if you can get close enough.

(3) Pay attention to the length of the bill as compared to overall body size, and note at the shape of the bill.

(4) Noting the length, shape, and color of legs and feet will prove useful.

(5) If you see the bird in flight, try to focus upon the color pattern of the wings, the color of the wing tips, the way the bird holds its head — whether tucked in or extended — the pattern of the flock, if any, and how often the bird flaps its wings. Some of these factors will vary with weather conditions, but as a general rule the manner in which a bird flies can go a long way in helping you toward positive identification.

Brown Pelican

Brown Pelican

The brown pelican (*Pelecanus erythorphthalmus*) is a species that has been in the news a great deal during recent years. Once quite prominent from the California coast all the way into South America, brown pelicans have been threatened to the point of becoming an endangered species because of the problems imposed by the industrialized world. Flocks of brown pelicans within the state produced thousands of young each year until the early 1960's when the population experienced a dramatic decline. Studies have shown the demise to be due to the accumulation of pesticides such as DDT which caused sterility and thin egg shells which broke while being naturally incubated. In 1970, it is believed that only 3 young were hatched throughout the entire state. Fortunately, the species has made a comeback, but their ultimate fate remains uncertain.

Brown pelicans can be identified by their general pelican shape, their long brown bill and throat pouch (gular pouch), and the overall brownish gray tone of their feathers. As adults, the birds often appear silvery above and have a conspicuous white or cream colored area on the head and neck which turns reddish brown during the breeding season of March through early August. Fully grown brown pelicans have a 6½ foot wingspan. These pelicans are most frequently seen in California during summer and fall. The adults do not make any noise.

When flying, but not actively fishing, brown pelicans are commonly seen within a few feet of the surface as they cruise along the windward

216

side of long ocean swells. They use the cushion of air for added lift. In flocks the birds almost always arrange themselves in orderly lines or in a well formed "V" formation.

When feeding, the normal fishing method is to fly along some 20 to 40 feet above the surface until the prey is located. Then the pelican makes a sharp downward plunge, while extending its neck, and holding its wings well back along the body, before making a large splash as the bird hits the water. Brown pelicans dive well below the surface to capture their prey. After the fish is caught, the bird returns to the surface to drain its pouch, during which time several species of gulls commonly referred to as thief birds are often seen trying to rob the brown pelican of its catch. The primary culprits are Heermann's gulls.

Brown pelicans can be easily distinguished from white pelicans, *Pelecanus erythrorhynchos*, by their coloration, size, and by the method of feeding. White pelicans, which when full grown are larger than brown pelicans, are more likely to be confused with snow geese than with brown pelicans. And when feeding, white pelicans do not dive below the surface of the water as brown pelicans often do. Instead, white pelicans often feed in groups of 3 or 4 birds that work surface fish into a tight circle before skimming the surface for their unfortunate prey.

Gulls

Several species of gulls can be seen along California beaches and at the islands throughout the year. Though they share many common characteristics, distinguishing between the most frequently observed species is not hard, as all species also have their own idiosynchracies. Gulls are generally considered to be medium sized birds whose flying abilities are average when compared to other birds.

The California Gull

California gulls, *Larus californicus*, inhabit a wide section of coastal North America from Southern California to British Columbia. Generally described as a medium-large gull as adults, the mantle is a neutral gray, while the outer primary feathers become increasingly black. White mirrors are evident on the last two primaries. Mirrors are brightly colored areas on darkly colored wings that are sometimes called a speculum. The legs and feet are a light grayish-green to yellow, the bill is yellow with a red dot, and the eyelids are edged in orangish red.

Along the coast, California gulls tend to scavenge, but when inland, they are known to follow farmers feeding on grubs that have been overturned by plowing. These birds also prey upon both insects and mice, when abundant. While commonly seen hundreds of miles inland, California gulls are not often seen more than 5 or 10 miles out to sea.

Western Gull

Western gulls, *Larus occidentails*, are abundant along the outer coast of California throughout the year. They are the largest of the California gulls, but they are among the easiest to confuse with other species. Westerns are, however, one of the few gulls to have all white heads as adults. The wings and back are dark, and the bill is yellow with a red spot on the mandible. And although not always easy to see, the feet of western gulls are pinkish. Foot color can be used to distinguish these birds from California gulls who have greenish feet.

Nesting in a low mound of soft plants on cliffs and at the offshore islands, western gulls are the only gulls to nest along the California coast. When nesting, the males establish a small territory which they constantly patrol by walking or just by sitting for hours on end. If another male enters the realm a fight will quickly ensue. The courtship ritual centers around the male's offering of a disgorged fish to an accepting female.

Western gulls often prey upon young cormorants and unguarded cormorant eggs.

Herring Gull

Being a very common and well-studied bird, herring gulls, *Larus argentatus*, are best known for two idiosyncracies displayed when feeding. Herring gulls quite commonly drop clams and mussels from a height of 40 to 50 feet onto hard rocks or packed sand in order to crack them open. They are also known to "puddle" or rapidly stomp in mud in order to drive worms to the surface for feeding.

Herring gulls are characterized by their pale gray mantle, relatively small black wing-tips with white mirrors on the outer primary feathers, white to lightly colored eyes with yellow irises, flesh colored legs, and large overall size.

Western Gull

Heermann's Gull

Heermann's Gull

Heermann's gulls, *Larus heermanni*, are medium sized gulls that are usually so dark in all stages of life that they are unlikely to be confused with other species. Adults have a dark gray mantle with even blacker wing-tips. The wing-tips lack a mirror, or speculum. The tail is black with a narrow white tip, and the head is mostly white though it can pick up some gray in fall and winter. The bills of Heermann's gulls are usually bright red, or red with a yellow tip, and the feet and legs are black.

Heermann's gulls are strictly associated with the outer seacoasts and nearshore waters. They drink salt water, and like many marine birds, these gulls secrete excess salt through two openings in their nostrils. Heermann's gulls are most commonly seen feeding in kelp forests, and are noted for gathering in a group around a fishing pelican in an attempt to steal the pelican's catch. After catching a fish, a pelican almost always returns to the surface to reposition its prey before swallowing. When they do so, Heermann's gulls frequently try to steal the fish by grabbing at its tail.

In Northern California, Heermann's gulls are often seen gathered around feeding sea otters, as they wait for the otters to cast aside small pieces of sea urchins and other prey. When fending for themselves, these gulls feed upon small fish, mollusks, shrimp, amphipods, and many types of dead animals.

Heermann's gulls are most often seen in Northern California during midsummer and fall, while commonly being seen further south throughout most of the year. However, the migratory pattern of this species is unusual in that a large part of the population travels in opposite directions from most birds. Heermann's gulls fly south to islands in the Sea of Cortez in the summer to breed, and then migrate north during winter.

Willets

Willets, *Catoptrophorus semipalmatus*, are the most widespread of the common large shorebirds in California. In addition, willets are the easiest of the large shorebirds to identify in flight due to the presence of 2 prominent, broad white bands and black stripes across their breasts. When a willet is at rest, the breast appears whitish and the back is gray. Willets feed predominantly upon a variety of shore crabs. They are often identified by their nervous, rather jerky manner of movement.

Long-Billed Curlew

The long-billed curlew, *Numenius americanus*, has the longest beak of any of the shorebirds. Often seen racing along California beaches as waves roll in, long-billed curlews are easily distinguished by their slender neck and long, downward curved bill. The body is a mottled brown. Because of its very long bill, these birds are able to feed upon polychaete worms, burrowing crustaceans, some small snails, and bivalve mollusks that are found deep in the mud along tidal flats.

Marbled Godwits

The marbled godwit, *Limosa fedosa*, is a shorebird that can be easily identified by its long, upturned bill. Shorter than long-billed curlews, marbled godwits reach a height of about 20 inches. Marbled godwits are colored a cinnamon brown, with highlighted darker markings covering the body. On sandy beaches which are exposed at low tide, flocks of marbled godwits are often seen scurrying along the beach as they hunt for small snails, worms, crabs, and amphipods.

Marbled godwits are often seen standing on one leg with their eyes shut and their bill tucked into their feathers with their head turned to the side. Marbled godwits are most common within California from August through April.

Willet

Long-billed
Curlew

Marbled
Godwit

Sanderling

Sanderling

Sanderlings, *Calidris alba*, frequent sand beaches all along the
California coast. About 8 inches in height, their plumage varies from
rusty gold during breeding season (April-May), to pale gray on top and
light below during the rest of the year. Sanderlings are a type of
sandpiper, and are a species that is commonly admired by
beachcombers in spring, summer, and early fall, as they scurry along the
line of waves at the water's edge. Sanderlings tend to feed in flocks,
searching for food in the backwash of large waves. Timing the outward
flow of the wave, the entire flock will rush in and probe the sand for the
younger stages of a burrowing sand crab, *Emerita analoga*. The
sanderlings hurriedly scamper out of the way of the next oncoming wave.

In calmer areas, sanderlings feed upon amphipods and other small
crustaceans in a more leisurely fashion. Showing a strong preference for
sandy beaches, sanderlings often gather together high up on beaches
when conditions are too rough for feeding, or in times of high winds.
Sanderlings travel to the Arctic tundra areas for breeding.

Elegant Tern

Common Tern

Terns

A number of species of terns are commonly seen in California. Among them are the Forster's tern (*Sterna forsteri*), the common tern (*Sterna hirundo*), the Arctic tern (*Sterna paradisaea*), the least tern (*Sterna albifrons*), and and the elegant tern (*Thalasseus elegans*). As a rule terns are especially graceful in flight. They must fly in order to obtain food, and as a result you can see terns on the wing in almost every kind of weather. Their wingbeats vary according to the wind conditions. When flying into a strong headwind, their wingbeats are rapid and deep, but in calmer winds even a slight downstroke of the wings produces a sharp upward response.

When feeding, terns hold their beaks at about a 45 degree angle below a horizontal plane, stopping abruptly when prey is sighted, and almost immediately beginning a steep vertical plunge into the water. The wings are folded during the dive. Small fish are the main source of food, but small crustaceans, amphipods, and insects are also taken. Various species are seen around bays, rocky outer coasts, and over deep oceanic waters.

It is interesting to note that Arctic terns breed over 11,000 miles from their wintering area, and thus must travel over 150 miles per day for twenty straight weeks in order to make the annual roundtrip migration. The figure of 11,000 miles represents the shortest distance possible, while in actual fact, field studies have documented that some Arctic terns travel from North America to Europe before heading south to Africa. Arctic terns are not as common as many other terns in California, and are generally seen in flight over oceanic waters. These birds breed in Arctic and sub-Arctic areas, while wintering in Antarctic and sub-Antarctic seas.

Great Blue Heron

Snowy Egret

The Great Blue Heron

The great blue heron, *Ardea herodias*, is easily distinguished from all other birds in California, except perhaps the sandhill crane, by its large size, long legs, gray coloration, and long neck. In flight, herons tuck their necks, while cranes extend their necks far forward. Adult great blue herons have a head that is mostly white, with a black stripe that extends into a plume on the top of the head. Cranes lack plumes. The body feathers are blue-gray. When standing alert, great blue herons are close to 4 feet tall.

Great blue herons prefer to feed upon fish, but are also known to take crustaceans, small reptiles, insects, and some rodents. The birds are usually solitary when hunting, slowly stalking along until they make a quick lunge for their prey. Great blue herons can often be seen near the marshes when hiking at Catalina Island.

Snowy Egret

Snowy egrets (*Egretta thula*) are the most common of the all-white members of the family of herons, bitterns, and egrets. They are characterized by their black legs and yellow feet. Often heard to make a low croaking sound, snowy egrets prey primarily upon small fish, a variety of crustaceans, and some large insects. Agile feeders, these birds often rush around in water in order to stir up food with their feet.

Once hunted to near extinction for their breeding plumage which was used to decorate hats, these birds have made a remarkable recovery. Snowy egrets are now commonly observed in many areas. Snowy egrets are often seen by those who hike along the marshes.

Pink-footed Shearwater

Sooty Shearwater

Sooty Shearwater

Shearwaters have a gull-like shape, but are generally smaller, and have shorter bills. Most are described as sooty brown. The bill and feet are black. Sooty shearwaters (*Puffinus griseus*) are all brown, except for variable amounts of silvery gray on the underwings. In California, sooty shearwaters are most commonly seen in spring, summer, and early fall over open water several miles offshore. These birds breed on islands near Australia, New Zealand, and South America.

The sooty shearwater is the only species of shearwater that occurs in flocks of thousands, a characteristic that aids greatly in positive identification. Some flocks have been estimated to contain over 500,000 birds. When gathered in flocks, the birds are usually feeding. Interestingly, it has been noted that the flocks often disperse within hours after feeding has been completed, but that is not always the case. Sooty shearwaters feed upon crabs, crab larvae, squid, and small schooling fishes such as anchovies.

Pink-Footed Shearwater

Like sooty shearwaters, pink-footed shearwaters, *Puffinus creatopus*, are perennial visitors throughout the length of the state. Their habitat is the open sea, and they are most numerous during the summer. While often mixing with flocks of sooty shearwaters, pink-footeds do so at a ratio of less than 100 to 1. Pink footed shearwaters are larger than sooties, have whitish undersides, and all pink feet. Pink-footed shearwaters breed on islands off the coast of Chile.

224

Pelagic Cormorant

Western Grebe

Black Oystercatcher

Black Turnstone

Western Grebe

The largest of grebes, the western grebe, *Aechmophotus occidentalis*, has a black upper body, a grayish mid-section, a white lower surface, and a long, thin neck. The beak is light yellow and quite straight. Grebes look a great deal like ducks, but differ due to a lack of distinct tail feathers.

During winter, these birds often gather in large groups near kelp beds. In the large "rafts" of birds, many sleep, making their long necks less obvious. When feeding, western grebes, like many other grebes, make dives for small fish which they spear with their sharp beaks. These birds make extended dives, being able to remain submerged for longer than one minute. In addition to fish, western grebes prey upon a few select species of crustaceans and other invertebrates.

Black Oystercatcher

Black oystercatchers, *Haematopus bachmani*, are seen along rocky shores at the offshore islands, as well as along the mainland. Usually described as a large shorebird, black oystercatchers reach the size of a small duck. They have all black feathering, a long, bright red beak, and pink legs. Using their sharp, stout beaks to pry off and crack open prey, these birds most often feed upon mussels, limpets, and barnacles that are found on rocky shores. Black oystercatchers are more common in the northern part of the state. On the mainland, they breed north of Morro Bay, but at the islands are known to nest as far south as the Coronados Islands off the coast of Tijuana, Mexico.

Cormorants

Three species of cormorants inhabit California. They are the pelagic cormorant (*Phalacrocorax pelagicus*), Brandt's cormorant (*Phalacrocorax penicillatus*), and the double-crested cormorant (*Phalacrocorax auritus*). Of these, the pelagic cormorants are the most common throughout the year. Like all cormorants, pelagics have fully webbed feet. This species is smaller than the other two, being about the size of an average duck, as opposed to the goose-sized bodies of the other cormorants. Distinctive features of pelagic cormorants include their slender neck and small head. When in flight, flocks often fly in long, straight lines, cruising just above the water, and then flapping their wings heavily.

Cormorants are often seen underwater by divers, as they seek their diet of fish and crustaceans. Sightings are well documented as deep as 80 feet, and have been reported at 120 feet.

Black Turnstone

When bird watching in rocky regions along the coast, quiet observers will often see black turnstones, *Arenaria melanocephala*. They are usually very active as they are constantly turning over rocks and clumps of seaweed in their constant quest for food. Black turnstones are similar in size to blue jays. Their dark plumage and white underbelly help them blend into their environment when they are on rocky beaches. Black turnstones are slightly larger than most sandpipers and sanderlings.

226

Indexes

Marine Biology Glossary
Cross-Reference Index
Selected Bibliography
Index

Glossary

The purpose of this glossary is to provide a quick reference source for some terms that are often used when describing the marine environment or the plants and animals found within. The following definitions are given for practical purposes, and are not intended as the definitive, all-inclusive scientific definitions. Many of the terms have more than one meaning. The definitions included in this glossary are intended to help readers with the use as it is likely to apply to topics discussed in this book. This glossary will also prove useful if you choose to refer to strict scientific texts for additional information.

A

aboral — away from the mouth or head.

abyssal realm — that portion of the marine environment in which light does not significantly penetrate. This region includes all water below a depth of 600 feet.

agamete — a single cell other than a gamete which is capable of developing into a complete new organism. In the case of plants, an agamete is a cell produced during the asexual generation of the cycle of alternation of generations.

agamic — asexual reproduction.

air bladder — swim bladder in bony fishes. In plants such as kelp, the air bladder is a gas-filled chamber that buoys the plant.

air sac — alveolar chamber in the lungs.

algin — natural by-product found in giant kelp and other brown seaweeds. Algin has a high affinity for water and is used as a thickening, emulsifying, and gel producing agent.

alternation of generations — the alternation of a sexually reproducing generation and an asexually reproducing generation in the reproduction of non-flowering plants.

ambergris — substance found in the intestinal tract of sperm whales, often used in perfumes.

amphibian — a cold-blooded invertebrate that lives in water and breathes by gills in the larval stage and that breathes by lungs as an adult. Amphibians are capable of living both in water and on land.

Ampullae of Lorenzini — sensory organs in the snout of sharks which help them detect either temperature fluctuations, pressure changes, or perhaps both.

anestrus — in female mammals the period of sexual inactivity between periods of heat.

angiosperm — any flowering plant.

annelid — member of the Phylum Annelida. Annelid worms are known as segmented worms.

anterior — toward the forward end or head.

anthozoan — any member of the class Anthozoa including sea anemones, jellyfish, and corals.

aperture — an orifice, as in the opening in the shell of abalone.

aphotic — without light. The aphotic zone of the oceans is that portion that receives little or no light. This region is considered to include all waters below 800 meters.

armored — having a protective covering of scales or bony plates.

arthropod — any invertebrate described in the phylum Arthropoda which includes lobsters, crabs, shrimp, isopods, amphipods, copepods, and other crustaceans.

ascidian — a bottom dwelling tunicate or sea squirt.

asexual reproduction — reproduction which does not involve the union of sperm and egg. Sporulation, budding, fragmentation, and fission are types of asexual reproduction.

attenuated shape — long and thin with a gradual taper.

autozoid — a feeding polyp found in some colonial anthozoans such as sea pansies. See siphonozooid.

B

baleen — flexible horny substance that grows from the upper jaw of filter feeding whales. Baleen is used as a sieve-like filter to trap food.

basal metabolism — minimum energy requirements of an organism in order to maintain normal bodily functions.

bathyl — of or having to do with the deeper regions of the oceans.

benthic — bottom dwelling. Benthic animals live on or near the bottom as opposed to pelagic organisms which live up in the water column.

binomial nomenclature — a naming sytem for organisms in which each life form is given a scientific name consisting of two words, the first designates the genus and the second designates the species. When correctly printed both the genus and species are italicised, and the genus is capitalized.

biological clock — an innate physiological rhythm that is often synchronized with environmental factors such as the rising or setting of the sun or the cycle of tides.

bioluminescence — production of light by living organisms to be distinguished from phosphorescence with which it is often confused by laymen. See phosphorescence.

biomass — the total mass of organic matter per unit of area. The term biomass is often used when explaining how much food is required to support predators.

biome — a community or zone that can be distinctively characterized by the plants or animals that live within.

bisexual — in some applications the term bisexual indicates that an animal possesses both male and female sexual organs.

bivalves — mollusks in the class Pelecypoda that have a shell which consists of two halves.

bladder — a sac in animals used as a reservoir for a gas or fluid. In plants, a bladder is a gas-containing chamber that is often called a pneumatocyst. The bladder buoys the plant.

blade — leaf-like appendage of a plant

blowhole — nostril or spiracle on the top of the head of whales and dolphins through which respiration occurs.

C

calcerous — containing calcium carbonate.

calciferous — creating or having calcite or carbonate of lime.

carapace — part of exoskeleton which covers the cephalothorax of some arthropods.

carnivore — an animal that devours the flesh of other animals, as opposed to plant eating herbivores.

caudal fin — tail fin.

cephalopod — any mollusk in the class Cephalopoda (head-footed) including octopi and squids.

chromatophore — a pigment cell used to alter colors in octopi, squids, some crustaceans, and other animals.

cilia — minute hairlike processes found along the edge of a cell. Cilia beat regularly and are used in locomotion.

clasper — structure used by males of all cartilaginous fishes during copulation. Normal males have a pair of claspers which are located on the underbelly.

class — a taxonomic category below a phylum and above an order.

classification — the systematic scientific categorization of plants and animals according to commonly shared characteristics.

cleaner — an organism that removes parasites, dead tissue, bacteria, or fungi from the surface of another animal.

cnidoblast — stinging cell found only in Cnidarians such as jellyfish and sea anemones which produces a nematocyst.

coelenterate — any invertebrate in the phylum Cnidaria including sea anemones, jellyfish, and corals.

cold blooded — having a core temperature which varies with ambient temperature. Most fishes, amphibians, and reptiles are considered to be cold blooded. Poikilothermous.

cold biological light — light produced by bioluminescent life forms.

colony — a group of living organisms that share a common skeletal case or test.

comb — a row of cilia found in ctenophores

commensalism — a type of symbiotic relationship in which the symbiont benefits but the host does not. The host is not harmed, but it does not benefit.

coralline — 1) an animal that bears strong resemblance to coral. 2) A type of red algae containing lime. These algaes are often important components of coral reefs.

crayfish — a freshwater crustacean that is closely related to lobsters. Most crayfish are smaller than most lobsters.

crustacean — any arthropod in the class Crustacea which includes lobsters, crabs, shrimps, copepods, isopods, and amphipods.

ctenophore — any member of the phylum Ctenophora which includes comb jellies.

D

decapod — any crustacean classified in the order Decapoda including lobsters, crabs, shrimps, amphipods, copepods, and isopods.

delayed implantation — phenomenon in which a fertilized egg remains unplanted in the uterine wall for some extended period of time, making the gestation period of shorter duration than the time between copulation and birthing. In some marine mammals with a gestation period of less than one year, delayed implantation helps to allow females to give birth and breed within a short period of time.

dermal denticle — tooth-like projections which comprise the skin of cartilaginous fishes. Sharks, rays, and skates lack scales, but do possess dermal denticles. The denticles align in one direction making sharks' skin feel very smooth if rubbed in one direction, but very coarse if rubbed in an opposing direction.

desiccation — eliminating or depriving of moisture, a major threat to animals that live in tide pool habitats.

devilfish — vernacular or common name for rays in the genus Manta. The term manta ray has replaced the term devilfish in recent years. Manta rays were believed to be dangerous by sailors in years past, hence the name devilfish.

diatoms — type of algae described in the class Bacillariophycae.

dinoflagellate — a protozoan described in the order Dinoflagellata having two flagella.

disc — a flat circular structure used to describe the colonial body of sea pansies and the central section of brittle stars.

distribution — the range or area that a species normally inhabits.

diurnal — pertaining to the day as opposed to nocturnal. Occuring daily.

dorsal — of or having to do with the back or upper surface of the body of an animal.

E

ecdysis — the process of molting in arthropods, in which an arthropod grows by discarding its exoskeleton, experiencing a period of rapid growth, and then forming a new, hard exoskeleton.

echolocation — biological sonar used by many species of mammals in which the animals send out a series of sounds which are reflected back to the sender off of other bodies, and are then analyzed by the sender to provide information concerning the size, speed, type of object etc.

echinoderm — any marine invertebrate described in the phylum Echinodermata which includes sea stars, brittle stars, sea cucumbers, crinoids, and sea urchins.

ecology — portion of the discipline of biology concerned with the interrelationships between organisms and between organisms and their surrounding environment, also known as environmental biology.

ecosystem — a unit found in nature consisting of all living and non-living forms—tide pools, kelp forests, and rivers for example.

ectoparasite — a parasite that lives on the skin of another animal, as opposed to an entozoan which is a parasite that lives within another organism.

ectozoan — an ectoparasite. See epizoan.

elasmobranch — any fish that is a member of the class Chondrichthyes, the cartilaginous fishes which includes sharks, skates, and rays.

entozoan — a parasite that lives within another organism, as opposed to an ectoparasite which is a parasite that lives on the surface of another animal.

estrus — period of sexual receptiveness in adult female mammals, also known as the period of heat.

epifauna — bottom dwelling animals that live on the surface of bottom materials.

epizoan — an animal which lives on the surface of another organism. See ectozoan.

estuary — a body of water where the end of a fresh water river mixes with sea water.

evolution — process by which organisms develop over extended periods of time as a result of changes and adaptations

excurrent — outflowing. In sponges the excurrent siphon helps to eliminate wastes and unwanted water, as opposed to the incurrent siphon which takes in water.

exoskeleton — a protective external skeleton as is found in lobsters, crabs, and other arthropods.

F

family — a taxonomic category, between order and genus.

fingerling — a fish as described between the time of the disappearance of the yolk sac and the end of one year.

finlet — a small fin located on either the dorsal or ventral sides near the tail of tuna and other fishes. Finlets serve to reduce drag.

flatworm — any member of the phylum Platyhelminthes, some of which are marine worms.

flower — a reproductive structure found in some plants.

fluke — a horizontal lobe on the tail of a whale. In casual conversation the term tail and fluke are sometimes used interchangeably.

food chain — a grouping of organisms in which energy is transferred from one trophic level to the next to the next etc. While microscopic plants form the foundation of many food chains, the chain often extends through large apex predators.

food web — a group of interlocking food chains.

fusiform — tapered at both ends. Fishes such as giant barracuda, albacore, and blue sharks are said to have a fusiform shape.

G

gamete — any cell which is capable of developing into a complete individual upon union with another sex cell.

gametophyte — a plant which produces gametes. In plants that reproduce in the method described as alternation of generations, the word gametophyte describes the gamete producing generation or plant.

gamic — sexual.

gastropod — any mollusk described in the class Gastropoda including abalone, other snails, limpets, periwinkles, and whelks.

genus — a taxonomic category between family and species. The genus name is always capitalized and should also be italicized.

gestation period — the period of time between the implantation of a fertilized egg in the uterine wall and birth.

gill — respiratory structure in aquatic animals through which gaseous exchange occurs.

gill rakers — projections located on the gill arches of some fishes which serve to prevent food particles from passing through the gill slits.

gill slits — one of several openings in the wall of the pharynx. In marine animals, the gill slits are separated by arches which bear gills.

girdle — the mantle of a chiton, or in vertebrates a structure that serves to support a part of the body such as a pelvic girdle.

gorgonian — corals described in the order Gorgonacea which includes sea fans. The skeletal case of gorgonians is comprised in part of a horn-like substance known as gorgonian.

H

habitat — a natural area such as a tide pool, kelp forest, or river that is considered to be the home of many organisms. A biome.

halophile — an organism that must have salt in its natural environment in order to survive.

halophyte — a plant that can successfully survive in areas with high salt concentrations such as lagoons and estuaries.

haptera — root-like looking structure which serves to attach a plant to the substrate. The haptera of a giant kelp plant collectively form a holdfast that secures the plant to the rocky bottom.

harem — female animals that mate with a single male, and some of whose behavior the male tries to control.

hermaphrodite — a plant or animal which at some time in its life possesses both male and female reproductive organs.

holdfast — structure of attachment or anchoring found in various marine plants. The holdfasts of giant kelp plants consist of numerous haptera. The holdfast secures the kelp plant to the bottom.

homocercal tail — a symmetrical or nearly symmetrical tail. A tail in which the upper and lower lobes are symmetrical or nearly so. A lunate tail. Mako sharks, great white sharks, and tuna have homocercal tails.

heterocercal tail — an asymmetrical tail in which the upper lobe is usually significantly larger than the lower lobe like the tail structure found in blue sharks and many other shark species.

I

icthyology — science that deals with the study of fishes.

implantation — in mammals, the process through which fertilized eggs become imbedded in the uterine wall.

indigenous — native to or naturally found in an area.

ink sac — a rectal gland found in cephalopods that serves to produce and store ink.

invertebrate — an animal that does not possess a backbone or spinal column.

iridocytes — specialized cells found in the skin of flatfish which help enable the fish to closely match the color and pattern of their skin with that of their surroundings.

isopod — any crustacean described in the order Isopoda, often observed on fish.

J

juvenile — a young or sexually immature organism.

K

kelp — any of several large brown seaweeds described in the order Laminariales.

L

lamprey — a jawless fish described in the class Marsipobranchii. Lampreys prey upon other fishes by sucking their blood and other body fluids.

larva — immature form of an organism that has been born but is unlike the adult form.

lateral line system — a series of sense organs in aquatic vertebrates which extends from the head to the tail along the sides of the body.

life cycle — the complete life history of an organism.

life history — the complete series of events displayed by an organism encompassing every stage between its origin and death.

littoral — of or having to do with the sea shore. The littoral area is the region between the tide lines. The term sublittoral describes the area below the low tide mark.

longitudinal — lengthwise or extending along the axis. Markings which run longitudinally run up and down the body, as opposed to across the body.

luminescence — production and emission of light without the accompanying production and emission of a significant amount of heat. Bioluminescent organisms create luminescent light.

M

macroplankton — planktonic organisms that can be recognized by the naked eye without any magnification.

mammal — any vertebrate described in the class Mammalia including man, whales, sea lions, seals, otters, seals, dolphins, and manatees. Mammals are characterized by hair and mammary glands which produce milk to feed offspring. All mammals breath air, have lungs, are warmblooded, and most bear live young.

mandible — in general usage, the jaw. In arthropods, one of a pair of parts in the mouth used to cut, crush, or grind food.

margin — the edge or border.

metabolic rate — the rate of chemical or energy changes which occur within a living organism as calculated by the amount of food consumed, the heat produced, or oxygen used.

metabolism — chemical and energy changes which take place within a living organism due to the activities involved in being alive.

metamorphosis — a process in which animals such as crustaceans undergo a change in shape or form as the animal develops from a fertilized egg to an adult.

migration — mass movement of populations of animals to and from feeding, breeding, and nesting areas.

milt — seminal fluid produced by the testes of fishes.

mirror — a brightly colored portion on the wings of some birds.

mollusk — any vertebrate described in the phylum Mollusca which includes a wide variety of specimens such as chitons, abalone, limpets, octopi, squids, nudibranchs, sea hares, clams, mussels, and oyster.

molt — the complete cycle of shedding and developing a new exoskeleton or outer covering.

mother-of-pearl — the inner layer of the shell of many mollusks such as abalone.

mussel — any bivalve mollusk described in the class Pelecypoda.

mutualism — a type of symbiotic relationship in which two organisms of differing species live in close association with each other to the advantage of both organisms. The relationship between moray eels and cleaner shrimp is described as mutualism.

N

nannoplankton — planktonic organisms less than 40 microns in length. A micron is one-thousandth of a millimeter, so 40 microns in 40/1000 of a millimeter, or .0016 inches.

narcosis — a state of confusion or stupor induced by a drug or foreign element. In the case of diving, the term nitrogen narcosis refers to a feeling that ranges from confusion to euphoria due to increased partial pressure of nitrogen in the human system at depth.

natural history — the study of or description of the life of various organisms. A complete discussion of the natural history of an organism would include information concerning the classification, habits, predator/prey relationships, life cycle, and distribution.

natural selection — process through which natural events determine which life forms will survive and which will perish.

nematocyst — in cnidarians, a specialized cellular capsule which contains a stinging nettle that can be used to paralyze prey. Not to be confused with the pneumatocyst found in plants. See pneumatocyst. Nematocysts serve in acquiring food, and as protective and adhesive structures.

neritic — being present in waters over a continental shelf. In general terms, the neritic zone includes all waters from the low tide zone to 600 feet.

nettle cell — stinging cell found inside the nematocysts of cnidarians.

notochord — in chordates, a dorsal rod of cartilage that runs the length of the body and forms the primitive axial skeleton in the embryonic stage of all chordates. In most adult chordates, the notochord is replaced by the spinal column. In tunicates the notochord forms the axial skeleton. In some respects a notocord is a precursor to a spinal column.

O

oozoid — in tunicates, an individual organism that develops from a mature egg cell or gamete.

operculum — a plate that covers the gill openings in some fishes.

order — a taxonomic category below a class and above a family.

organic — having to do with or having been created by living things.

organic evolution — process by which various organisms develop, change, and adapt over time.

osculum — the excurrent or outflowing opening in a sponge.

osmoregulator — an organism that maintains a constant concentration of various salts within the body in spite of the concentrations of the salts in the organism's immediate surrounding.

osmosis — passage of water through a tissue or membrane as a result of different concentrations of salts.

oviparous — a specific type of reproduction in which females lay eggs that hatch outside the body of the female. Compare with ovoviviparous and viviparous.

ovoviviparous — a specific type of reproduction in which females produce eggs encased in a shell which develop inside the body, but the young only receive nourishment from the yolk sac, and not directly from the mother.

P

paleontology — the study of plants and animals that lived during previous time periods according to what can be learned from fossil remains.

parasite — an organism which lives on or in another organism and from which it takes some nourishment.

parasitism — type of symbiotic relationship in which one organism, the parasite, is dependent at least in part on the other organism, the host, for food.

pectoral — of or having to do with the chest or breast area of a body.

pectoral fin — one of the paired, laterally oriented, fins of a fish. Pectoral fins are generally forward on the mid-body of fish.

pectoral girdle — the cartilaginous or bony structure that provides support for the pectoral fins of a fish.

pedicellaria — small pliers-like looking organs found on the body surface of some echinoderms such as sea stars. Under a microscope pedicellaria look like miniature pliers that are capable of pinching together. Pedicellaria serve to clean debris off the body surface.

peduncle — the stalk of a barnacle.

pelagic — of or having to do with the open ocean. Pelagic organisms live up in the water column as opposed to on the bottom. See benthic.

pelecypod — any mollusk described in the class Pelecypoda which includes mussels, scallops, clams, and oysters.

pharyngeal slit — in members of the phylum Chordata, one of a series of openings between the throat (or pharynx) and the surrounding environment.

pheromones — chemical substances used for communication between members of the same species. Pheromones are used to induce the simultaneous release of sex cells in many invertebrates that reproduce via external fertilization.

phoresis — a type of symbiotic relationship in which the host provides transportation for the symbiont. The relationship between barnacles and California gray whales is correctly described as phoresis.

phosphorescence — the production and emission of light without the accompanying production and emission of heat.

photic — having to do with light.

photic zone — that region in the marine realm where sunlight is able to penetrate in an amount sufficient to support photosynthesis.

photosynthesis — process through which plants convert radiant or solar energy into chemical energy that is stored in the molecules of various carbohydrates.

phylogenetic — having to do with ancestral development.

phylum — a taxonomic category comprising the largest category of the plant or animal kingdom. A phylum is subdivided into classes.

pinniped — any aquatic mammal described in the class Pinnipedia which includes seals and sea lions. Pinnipeds have modified limbs that are used as flippers.

placoid cell — a type of scale made of a bony plate that is embedded in the outer layer of skin or in the epidermis of elasmobranchs. Same as a dermal denticle.

plankton — free floating aquatic life forms that tend to float passively, having only limited control over their movement.

pneumatocyst — the air bladder or float found on various marine algaes. In kelp, pneumatocysts serve to buoy the plants toward the surface and sunlight.

pod — a group of whales, porpoises, or dolphins.

poikilothermous — cold-blooded; having a core temperature that varies with the surrounding environment.

polygamous — having more than one mate during a short period of time.

polyp — a small, bottom dwelling form of various coelenterates, animals described in the phylum Cnidaria, which are attached at the base and have a mouth that is surrounded by tentacles. Colonial animals are composed of groups of polyps that live within the same skeletal case.

posterior — at or towards the tail or back end of the body.

predation — the act of seeking out and capturing other animals for food.

predator — an animal which practices predation.

prey — an animal that is captured and eaten by another animal.

proboscis — the elongated snout of various animals such as that of an elephant or male elephant seal.

process — 1) a series of interconnected activities. 2) a projecting outgrowth on a body.

R

radula — platelike structure that bears rows of tiny teeth. A radula is found in many mollusks and is used as a tool for rasping and chewing.

regeneration — the replacement of lost tissues or body parts in higher life forms, or the development into a functional individual from a part other than a reproductive cell in lower animals such as sea stars.

respiration — the exchange of gases between an organism and its surroundings.

rostrum — a body projection extending forward from the head in manta rays and other animals.

S

sagittal crest — a pronounced arrow-shaped ridge found in the mid-line of the skull in certain marine mammals.

salp — common name for several types of free swimming marine tunicates described in the class Thaliacea.

scientific name — the taxonomic classification known as the genus and species of a given life form. The genus name should always be capitalized and both words should appear in italics.

sea star — a starfish, term that is replacing the word starfish in some educational circles.

seawater — saline water found in ocean basins. Salt contents vary, but average close to 3.5%.

seaweed — any of several large plants that grow in seawater including kelp

sediment — matter which settles to the bottom.

sexual generation — in plants that reproduce in a cycle known as alternation of generations, it is the generation which produces gametes, either eggs or sperm.

shorebird — common name for any of a large number of birds that frequent sand beaches, backwaters, bays, and estuaries. These birds usually have long legs and long beaks. Many are described in the suborder Charadrii.

siliceous — of or having to do with silicon dioxide, also known as silica, which is the basis of sand. Silica is present in approximately 27% of the earth's crust.

siphon — a tubelike structure used to draw in or expel fluids. Siphons are found in many bivalves, cephalopods, and tunicates.

siphonozooid — a modified polyp which serves to create water currents in some colonial anthozoans. See autozooids.

speciation — the process by which a new species is naturally formed.

speculum — brightly colored portion on the wings of some birds.

spermaceti — wax-like substance found in the head of sperm whales.

spermatophore — a packet of spermatozoa which is transferred to a female in various crustaceans, octopi, and other animlas.

spiracle — one of a pair of openings on the top of the head of rays, sharks, and skates, through which water is drawn in before passing over the gills. The term is also used to describe the blowhole of a cetacean.

splash zone — area above the high tide line that is often moistened with spray from breaking surf but is only rarely, if ever, completely awash.

stipe — a stalk or supporting structure as in the stipe of a kelp plant.

stony coral — a coelenterate described in the order Madreporaria. In California waters solitary corals are often referred to as stony corals.

subspecies — a subdivion of a species in which interbreeding is possible, but in which slight physiological differences are present.

substrate — solid material on which organisms live; the bottom or seafloor.

swimmeret — a slender, branched abdominal appendage found on the abdomen of crayfishes and lobsters. In females, the eggs are carried on the swimmerets.

symbiont — any organism living in a symbiotic relationship, or in symbiosis.

symbiosis — a life style in which two organisms of different species live in close association with each other. Symbiotic relationships are further defined as mutualism, commensalism, parasitism, or phoresis.

T

taxon — a taxonomic classification of any size such as a phylum, order, family, or species.

taxonomy — the arrangement and classification of plants and animals into categories based on commonly shared characteristics.

teleost — any fish with a bony skeleton, a member of the class Oestichthyes.

tentacle — any of a number of long, thin, flexible, unsegmented appendages that serve in a sensory, food getting, locomotive, defensive, attaching capacity, or reproductive capacity.

terminal — of or having to do with the front or forming end of a body. The mouth of a garibaldi is located in a terminal position, while the mouth of a blue shark is not.

territory — an area to which animals normally confine their activities. In many cases the animals vigorously defend the area from intruders.

threshold — the lowest limit at which a given event will occur. A survival threshold describes the limit to which an animal or species can be pushed before dying.

tide — the periodic, cyclical, and predictable rise and fall of the sea along along the coast.

trinomial nomenclature — an extension of the binomial system in which a subspecies or a variety is scientifically named with a total of three words, two of which combine to comprise the species name. When properly written all three words are italicised, but only the genus is capitalized.

trophic — of or having to do with growth or nutrition.

trophic level — a place in a food chain in which animals acquire their food from the same sources.

U

uniparous — bearing one offspring at a time.

univalve — a mollusk whose shell has only one valve, or plate that makes the shell.

V

ventral — of or having to do with the lower surface or underside of a body.

vernacular name — the common or non-scientific name of an organism.

vertebrate — any animal having a spinal column which is described in the subphylum Vertebrata.

viviparous — a form of reproduction in which the eggs of undeveloped young hatch inside the body of the mother, and obtain further nourishment from her as development proceeds. The young are born at a later point in time when development has been completed.

visceral mass — the main portion of the body of an abalone or bivalve mollusk which is located above the foot and includes the internal organs.

W

whorl — a spiral turn or twist in the shell of a gastropod mollusk.

X

xenology — the study of the relationships between animals that are hosts and parasites.

Z

zooid — an individual organism in a colony or closely associated group of animals.

zooplankton — animal plankton, as opposed to plant plankton.

Cross Reference Index from Common Name to Scientific Name

A

B

C

California halibut	*Paralichthys californicus*, page 159
California hydrocoral	*Allopora californica*, page 78
California moray eel	*Gymnothorax mordax*, page 122
California purple striped jellyfish	*Pelagia panopyra*, page 168
California sea hare	*Aplysia californica*, page 93
California sea mussel	*Mytilus californianus*, page 49
California sea lion	*Zalophus californianus*, page 198
California spiny lobster	*Panularus interuptus*, page 99
Catalina goby	*Lythrypnus dalli*, page 136
Catalina perch	*Medialuna californiensis*, page 131
checkered periwinkle	*Littorina scutulata*, page 50
chestnut cowrie	*Zonaria spadicea*, page 94
China rockfish	*Sebastes nebulosus*, page 124
cleaner shrimp	*Hippolysmata californica*, page 102
common dolphin	*Delphinus delphi*, page 193
common sand crab	*Emerita analoga*, page 54
common sea cucumber	*Stichopus parvimensis*, page 108
common squid	*Loligo opalescens*, page 147
common tern	*Sterna hirundo*, page 222
common thresher shark	*Alopias vulpinus*, page 177
conspicuous chiton	*Stenoplax conspicua*, page 49
coon stripe shrimp	*Pandalus danae*, page 102
corynactus anemone	*Corynactus californica*, page 77
C-O turbot	*Pleuronichthys coenosus*, page 160
crumb-of-bread sponge	*Halichondria panicea*, page 75

D

diamond turbot	*Hypsopetta guttulata*, page 160
double-crested cormorant	*Phalacrocorax auritus*, page 226

E

eel grass	*Zostera marina*, page 70
electric ray	*Torpedo californica*, page 154
elegant tern	*Thalasseus elegans*, page 222
elephant ear tunicate	*Polyclinum planum*, page 110
elephant seal	*Mirounga angustirostris*, page 205
elk kelp	*Pelagophycus porra*, page 60

F

feather-boa kelp	*Egregia leavigata*, page 59
feather-duster worm	*Eudistylia polymorpha*, page 82
file limpet	*Acmaea limatula*, page 90
finback whale	*Balaenoptera physalus*, page 192
fingered limpet	*Acmaea digitalis*, page 90
flat abalone	*Haliotus walallensis*, page 87
flat-bottomed periwinkle	*Littorina planaxis*, page 52
flat porcelain crab	*Petrolisthes cinctipes*, page 51
fleshy sea pen	*Leioptilus guerneyi*, page 143
fluted bryozoan	*Hippodiplosia pacifica*, page 145
Forster's tern	*Sterna forsteri*, page 222
fragile sea star	*Linckia columbiae*, page 104
fragile tube worm	*Salmacina tribanchiata*, page 82

G

gaper clam	*Tresus nuttalli*, page 55
garibaldi	*Hypsypops rubicundus*, page 133
giant green anemone	*Anthopleura xanthogrammica*, page 77
giant kelp	*Macrocystis pyrifera*, page 59
giant kelpfish	*Heterostichus rostratus*, page 66
giant key-hole limpet	*Megathura crenulata*, page 90
giant octopus	*Octopus dolfleini*, page 96
giant red urchin	*Strongylocentrotus franciscanus*, page 107
giant star or giant sea star	*Pisaster giganteus*, page 103
glassy tunicate	*Ascidia paratropa*, page 110
goose-neck barnacle	*Pollicipes polymerus*, page 98
gopher rockfish	*Sebastes carnatus*, page 124
gray moon sponge	*Spheciospongia confoederata*, page 75
gray puff ball sponge	*Tetilla arb*, page 76
great blue heron	*Ardea herodias*, page 223
great blue shark	*Prionace glauca*, page 172
great white shark	*Charcharodon carcharius*, page 115
green abalone	*Haliotis fulgens*, page 85
green sea urchin	*Strongylocentrotus drobachensis*, page 107
Guadalupe fur seal	*Arctocephalus townsendi*, page 207
guitarfish	*Rhinobatus productus*, page 156

H

hairy hermit crab	*Pagurus hirsutiusculus*, page 101
halfmoon	*Medialuna californiensis*, page 131
halibut	*Paralichthys californicus*, page 159
harbor seal	*Phoca vitulina*, page 204
heart urchin	*Lovenia cordiformis*, page 153
Heermann's gull	*Larus heermanni*, page 219
herring gull	*Larus argentus*, page 218
Hopkin's rose	*Hopkinsia rosacea*, page 92
horn shark	*Heterodontus francisci*, page 112
humpback whale	*Megaptera novaengliae*, page 192

I

island kelpfish	*Heterostichus rostratus*, page 127

J

jack mackerel	*Trachurus symmetricus*, page 129

K

Kellet's whelk	*Kelletia kelleti*, page 95
kelp bass	*Paralabrax clanthratus*, page 128
kelp clingfish	*Rimicola eigenmanni*, page 66
kelp limpet	*Acmaea insessa*, page 90
kelp surfperch	*Brachyistius frenatus*, page 66
key-hole limpet	*Megathura crenulata*, page 90
killer whale	*Orcinus orca*, page 196
knobby star or knobby sea star	*Pisaster giganteus*, page 103

243

L

lacy bryozoan	*Phidolopora pacifica*, page 145
laminaria kelp	*Laminaria dentigera*, page 60
large beach hopper	*Orchestoidea corniculata*, page 53
large coffee-bean cowrie	*Trivia solandri*, page 95
least tern	*Sterna albifrons*, page 222
leopard shark	*Triakia semifasciata*, page 114
Lewis' moon snail	*Polinices lewisii*, page 149
long-billed curlew	*Numenius americanus*, page 220
light bulb tunicate	*Clavelina huntsmani*, page 109
lingcod	*Ophiodon elongatus*, page 126
little coffee-bean cowrie	*Trivia californiana*, page 95
little-neck clam	*Protothaca staminea*, page 55

M

Macfarland's dorid	*Chromordis macfarlandi*, page 92
macrocystis kelp	*Macrocystis pyrifera*, page 59
mako shark (see shortfin mako shark)	
marbled godwit	*Limosa fedosa*, page 220
market squid	*Loligo opalescens*, page 147
matted green anemone	*Anthopleura elegantissima*, page 77
megamouth or megamouth shark	*Megachasma pelagios*, page 179
Metridium anemone	*Metridium senile*, page 77
minke whale	*Balaenoptera acutorostrata*, page 192
moray eel	*Gymnothorax mordax*, page 122
morning sun star	*Pycnopodia helianthoides*, page 104
mossy chiton	*Mopalia mucosa*, page 49
mouse urchin	*Lovenia cordiformis*, page 153
mussel blenny	*Hypsoblennius jenkinsi*, page 136

N

navanax	*Navanax inermis*, page 94
nickeleye goby	*Coryphoterus nicholsii*, page 136
Norris' top shell	*Norrisia norrisi*, page 64
northern elephant seal	*Mirounga angustirostris*, page 205
northern kelp crab	*Pugettia producta*, page 101
northern fur seal	*Callorhinus ursinus*, page 206

O

Obelia hydroids	*Oebelia sp.*, page 79
ocean whitefish	*Caulolatilus princeps*, page 129
ochre star or ochre sea star	*Pisaster ochraceus*, page 103
olive shells	*Olivella sp.*, page 95
opaleye	*Girella nigricans*, page 131
orange cup coral	*Balanophyllia elegans*, page 78
orange gorgonian	*Adelogoria phyllosciera*, page 80
orange sponge	*Tethya aurantia*, page 76
orange sun star	*Solaster stimpsoni*, page 104
orca	*Orcinus orca*, page 196
owl limpet	*Lottia gigantea*, page 90

P

Pacific mackerel	*Scomber japonicus*, page 180
Pacific sanddab	*Citharichthys sordidus*, page 160
painted greenling	*Oxylebius pictus*, page 126
palm kelp	*Eisenia arborea*, page 60
palm kelp	*Poftelsia palmaeformis*, page 60
palm kelp	*Pterygophora californica*, page 60
parasitic anemone	*Epizoanthus scotinus*, page 80
pea crab	*Pinnixa littoralis*, page 51
pelagic cormorant	*Phalacrocorax pelagicus*, page 226
pilot whale	*Globicephala macrorhyncus*, page 193
pink abalone	*Haliotus corrugata*, page 87
pink footed shearwater	*Puffinus creatopus*, page 224
pink sea pen	*Leioptilus guerneyi*, page 143
pinto abalone	*Haliotus kamtschatkana kamtschatkana*, page 87
pismo clam	*Tivela stultorum*, page 55
plume hydroid	*Plumaria alicia*, page 79
plume worm	*Serpula vermicularis*, page 82
puff ball sponge	*Tethya aurantia*, page 76
pugnacious aeolid	*Phidiana pugnax*, page 92
purple coral	*Allopora californica*, page 78
purple gorgonian	*Eugorgia rubens*, page 80
purple sea urchin	*Strongylocentrotus purpatus*, page 107

R

rainbow dendronotid	*Dendronotus iris*, page 92
ratfish	*Hydrolagus colliei*, page 122
red abalone	*Haliotus rufescens*, page 87
red crab	*Cancer productus*, page 101
red gorgonian	*Lophogorgia chilensis*, page 80
red Irish Lord	*Hemilepidotus hemilepidotus*, page 127
red sea star	*Mediaster aequalis*, page 104
red sea urchin	*Strongolycentrotus franciscanus*, page 107
red turban	*Astraea gibberosa*, page 95
ribbon kelp	*Egregia laevigata*, page 59
rock louse	*Ligia occidentalis*, page 48
rockpool blenny	*Hypsoblennius gentillis*, page 136
rock scallop	*Hinnetes multirugosus*, page 95
rock sole	*Lepidopsetta bilineata*, page 160
rock wrasse	*Halichoeres semicinctus*, page 135
rose star	*Crossaster papposus*, page 104
rosy rockfish	*Sebastes rosaceus*, page 124
rosy sculpin	*Oligocottus rubellio*, page 127
rubberlip surfperch	*Rhacochilus toxotes*, page 132

S

sand castle worm	*Phragmatopoma californica*, page 82
sand dollar	*Dendraster excentricus*, page 152
sanderling	*Calidris alba*, page 221

246

T

Taylor's sea hare	*Aplysia taylori,* page 98
Tealia anemone	*Tealia crassicornis,* page 77
Tealia anemone	*Tealia lofotensis,* page 77
thick-horned aeolid	*Hermissenda crassicornis,* page 92
thornback ray	*Playtyrhinoidis triseriata,* page 154
threaded abalone	*Haliotus kamtschatkana,* page 87
three-color polycera	*Polycera tricolor,* page 92
three wing murex	*Pteropura trialata,* page 95
tidepool sculpin	*Oligocottus maculosus,* page 127
treefish	*Sebastes serriceps,* page 124
troglodyte chiton	*Nuttallina.californica,* page 49
tube anemone	*Cerianthus aestuari,* page 143
two-spot or two-spotted octpus	*Octopus bimaculatus,* page 96

U

urn sponge	*Rhabdodermella nuttingi,* page 75

V

vanilla sponge	*Xestospongia vanilla,* page 76
vermilion rockfish	*Sebastes miniatus,* page 124
volcano limpet	*Fissurella volcano,* page 90

W

walleye surfperch	*Hyperprosopon genteum,* page 132
wart-necked piddock clam	*Chalae ovidae,* page 145
wavy cockle	*Chione unidatella,* page 55
wavy top turban	*Astraea undosa,* page 95
western grebe	*Aechmophotus occidentalis,* page 225
western gull	*Larus occidentalis,* page 218
white abalone	*Haliotus sorenseni,* page 87
white lined dorid	*Dirona albolineta,* page 92
white seabass	*Cynoscion noblis,* page 130
white sea cucumber	*Eupentacta Quinnuesemita,* page 108
white sea pen	*Stylatula elongata,* page 143
white sea urchin	*Lytechinus anamesus,* page 153
willet	*Catoptrophorus semipalmatus,* page 220
wolf eel	*Anarrhichthys ocellatus,* page 135
wolly sculpin	*Clinocottus analis,* page 127

Y

yellow sponge	*Mycale macginitie,* page 75
yellowtail	*Seriola dorsalis,* page 182

Z

zebra goby	*Lythrypnus zebra,* page 136

Selected Bibliography

The following selected bibliography is by no means an attempt to list all of the published books and articles that were consulted in preparing this text. Instead, and for the benefit of those readers interested in learning more about California marine life, I include here some of the books and reference material that I found to be particularly interesting and useful.

Allen, Richard K. 1980. *Common Intertidal Invertebrates of Southern California.* Peek Publications. Palo Alto, California.

Behrens, David W. 1980. *Pacific Coast Nudibranchs.* Sea Challengers. Los Osos, California.

Berger, Wolf. 1979. *Walk Along the Ocean.* Beach Walk. Solana Beach, California.

Burton, Carol M. and Paul C. 1983. *Tidepool Life.* Opportunities for Learning, Inc. Chatsworth, California.

Castro Jose I. 1983. *The Sharks of North American Waters.* Texas A&M University Press. College Station, Texas.

Cogswell, Howard L. 1977. *Water Birds of California.* University of California Press. Berkeley, Los Angeles, and London.

Coulombe, Deborah A. 1984. *The Seaside Naturalist.* Phalarope Books. New Hampshire.

Daugherty, Anita E. 1979. *Marine Mammals of California.* Sea Grant Marine Advisory Program. California.

Fitch, John E. and Lavenberg, Robert J. 1971. *California Marine Food and Game Fishes.* University of California Press. Berkeley, Los Angeles, and London.

Galbraith, Robert and Boehler, Ted. 1974. *Subtidal Marine Biology of California.* Naturegraph Publishers Inc. Happy Camp, California.

Gotshall, Daniel W. 1977. *Fishwatchers' Guide to the Inshore Fishes of the Pacific Coast.* Sea Challengers. Monterey, California.

Gotshall, Daniel W. 1982. *Marine Animals of Baja California.* Sea Challengers. Los Osos, California.

Gotshall, Daniel W. and Laurent, Laurence L. 1979. *Pacific Coast Subtidal Marine Invertebrates.* Sea Challengers. Los Osos, California.

249

Hedgepeth, Joel and Hinton, Sam. 1961. *Common Seashore Life of Southern California.* Naturegraph Publishers Inc. Happy Camp, California.

Hinton, Sam. 1969. *Seashore Life of Southern California.* University of California Press. Berkeley, Los Angeles, and London.

Howarth, Peter C. 1978. *The Abalone Book.* Naturegraph Publishers, Inc. Happy Camp California.

Leatherwood, Stephen and Reeves, Randall R. 1983. *The Sierra Club Handbook of Whales and Dolphins.* Sierra Club Books. San Francisco, California.

McCormick, Harold W. and Allen, Tom with Young, William E. 1978. *The Sharks, Skates and Rays, Shadows in the Sea.* Stein and Day Publishers. New York, New York.

McIntyre, Joan. 1974. *Mind In The Waters.* Charles Scribner's Sons. New York, New York.

Miller, Daniel J. and Lea, Robert N. 1976. *Guide to the Coastal Marine Fishes of California.* University of California. Richmond, California.

Miller, Tom. 1975. *The World of the California Gray Whale.* Baja Trail Publications, Inc. Santa Ana, California.

North, Wheeler J. 1976. *Underwater California.* University of California Press. Berkeley, Los Angeles, and London.

Peterson, Richard S. and Bartholomew, George A. 1967. *The Natural History and Behavior of the California Sea Lion.* American Society of Mammalologists. Okalahoma State University. Stillwater, Oklahoma.

Raven, Peter H. and Johnson, George B. 1986. *Biology.* Times Mirror/Mosby College Publishing. St. Louis, Toronto, Santa Clara.

Reisch, Donald J. 1972. *Marine Life of Southern California.* Forty-Niner Shops, Inc. Long Beach, California.

Steen, Edwin B. 1971. *Dictionary of Biology.* Barnes and Noble Books. New York, Hagerstown, San Francisco, London.

Sumich, James L. 1976. *Biology of Marine Life.* Wm. C. Brown Company Publishers. Dubuque, Iowa.

Walker, Theodore J. 1979. *Whale Primer.* Cabrillo Historical Association. San Diego, California.

Index